Teaching Biology Today

Teaching Biology Today

Dorothy Dallas

Hutchinson
London Melbourne Sydney Auckland Johannesburg

Hutchinson & Co. (Publishers) Ltd

An imprint of the Hutchinson Publishing Group

24 Highbury Crescent, London N5 1RX

Hutchinson Group (Australia) Pty Ltd
30–32 Cremorne Street, Richmond South, Victoria 3121
PO Box 151, Broadway, New South Wales 2007

Hutchinson Group (NZ) Ltd
32–34 View Road, PO Box 40-086, Glenfield, Auckland 10

Hutchinson Group (SA) (Pty) Ltd
PO Box 337, Bergvlei 2012, South Africa

First published 1980

© John Barker 1980

The paperback edition of this book is sold subject to the condition that it shall not, by way of trade or otherwise, be lent, resold, hired out, or otherwise circulated without the publisher's prior consent in any form of binding or cover other than that in which it is published and without a similar condition including this condition being imposed on the subsequent purchaser

Set in 10/12 Times Roman

Printed in Great Britain by The Anchor Press Ltd
and bound by Wm Brendon & Son Ltd
both of Tiptree, Essex

British Library Cataloguing in Publication Data

Dallas, Dorothy
  Teaching biology today.
  1. Biology – Study and teaching – Great Britain
  I. Title II. Barker, John
  574' .07'1041   QH320.G7

ISBN 0 09 141091 6 (paper)

This book is dedicated with love, respect and admiration to my two Nuffield friends,
    Hilda Misselbrook and Grace Monger

Si monumentum requiris, circumspice!

Contents

Preface ix

Acknowledgements x

1 Biology teaching — an introduction 1
2 Lesson planning 20
3 Mixed ability teaching 40
4 A variety of methods 66
5 Difficult or impossible to teach? 119
6 Biology's special needs 124

Appendices

A An example of the historical approach 136
B Worksheets used in team teaching 138
C Use of the clinical thermometer 147
D *Quest* worksheets: basic skills 151
E An independent learning approach for the sixth form 153
F Resource lists 155

References 167
Further information 175
Index 177

Preface

Dorothy Dallas died before she could see the publication of this book. It is the result of a lifetime's work in the field of biological education and her wish was that it would help to improve the quality of biology taught in our schools. Dorothy's style and mode of expression were something uniquely her own and I have done my best to retain them in the book. I hope this will provide a tangible means by which all her friends can remember her.

John A. Barker

Acknowledgements

To past students of the biology method course at King's College London, who have taught me a great deal. Some of their work appears in the text. Particular thanks are due to Denis Ayling, John Bilsborough, John Blackett, Peter Brown, Maggy Campbell, Martyn Chesworth, Ruth Genner, Don Gray, Jeremy Ison, Graham Jones, Karin Kay, Raza Khan, Sue Knight, Zoe Manning, Les Pickett, Delia Richards, Dave Wright, and anyone I've missed.

To John Barker, of the Centre for Science Education, Chelsea College, University of London, who wields a deft pruning knife.

To Brian Dudley, of the Department of Education, Keele University for a neat piece of hatchet work, greatly appreciated.

To Clare Smallman, for her initial editing of the text.

To the anonymous author of 'Giky Martables'. The work was retrieved from the floor at a VSO course by Clive Carré of Exeter University, and has delighted students for some years since.

To Mike Wild of Totley-Thornbridge College of Education, Sheffield for permission to publish his diagram on discussion arrangements.

To Longman and the Nuffield Foundation for permission to print the graphs and nomogram on page 51 and the crossword on page 106.

To Mrs Eldridge, for her expertise in interpreting my handwriting and for typing the text with persistence, determination and amazing accuracy.

To Thames Television for permission to publish the worksheets on pages 34–6.

The editor would like to thank the members of UBET (University Biology Education Tutors) who read and commented upon the final version, and Miss Pat Greenwood who typed it.

1 Biology teaching – an introduction

The collection code is capable of working when staffed by mediocre teachers, whereas integrated codes call for much greater powers of synthesis and analogy and far more ability. . . .

B. BERNSTEIN,
in M. F. D. Young (ed.), *Knowledge and Control* (Collier Macmillan 1971)

There is no one way of teaching biology, just as there is no definition of the factual content covered by the term; it can range from the exact sciences of physics, chemistry and maths, to the inexact and exciting realms of human behaviour. Some teachers get satisfaction by playing at professors, and sending their students to university having done all their first year's work; others get very bored teaching examination work to bright pupils who would probably have passed without much intervention from them. Bright pupils are satisfying to teach if there is something to argue about, but if the teacher has only to hand over facts many find it stultifying. On the other hand the sheer repetition and slow pace needed for the least able pupils can be equally stultifying. But remember that most of the school population lies somewhere between the two extremes, and that most of the time it is exciting, rewarding, frustrating *and* boring to teach them.

Biology defies classification as one particular form of knowledge as it must necessarily deal with all kinds of disciplines. Biologists must therefore prepare themselves to admit ignorance; no one can possibly have a sound grasp of all the relevant fields, and pupils will ask unanswerable questions. Teachers therefore need to be confident enough to admit what they don't know, and some find this difficult to reconcile with their assumed role as the fount of all knowledge for their pupils. Such a situation may make some pupils feel that the teacher is an idiot, but if the teacher proceeds to examine ways in which to find out the elusive answer, something of the excitement of science can begin. Unfortunately many teachers have themselves been taught by assimilating information and confirming experimental results, and after twenty-one years of conditioning many emerge from their professional training teaching in exactly the same way as they have been taught themselves. This is not always a bad thing, if they are teaching those whose socialization has been the same as their own, but within every culture and every country there are a variety of subcultures. The attitude of some of them to school may well be that it is irrelevant to their life and aspirations, and their home life may not provide the support necessary for the particular type of teaching associated with a traditional formal education.

It is wise therefore to look at methods of teaching other than those with which one was taught – not to jump uncritically on a gimmicky bandwagon, but to examine the use of various methods and their application by one's own personality to a particular set of pupils or even to individuals. A systematic approach to teaching and learning has been adopted by many universities, particularly in medical schools, yet it is still considered undignified in many secondary schools to use anything but formal methods. It is to these that many teachers return in situations of stress, punctuating their discourse by cries of 'Quiet', 'Listen', 'Sssh', and so on. This is not education.

Biology teaching has evolved from what could be called necrology (the study of lots of grey pickled things, not forgetting the ten cranial nerves of the dogfish) to an emphasis on first-hand evidence; the dissection of a fresh dogfish is not only aesthetically more pleasing, but vastly more instructive. Today living things are seen much more in relation to each other and their habitat. Also there has been a swing of the pendulum away from chalk and talk — somewhat uncritically, perhaps, as there are times when this is the right and relevant method. There has been a swing towards the uncritical belief that if pupils are doing practical work, they are also learning. In many practical situations it is technology which is being learnt and manual dexterity which is being practised; at worst practical work can be purely recipe-following without comprehension. New teachers tend to show their attitudes to practical work with the well-worn phrase 'It will fill up the time nicely' or 'But how can I teach it, there's no practical?' A systematic approach, listing what the pupils need to learn and the best method of helping them to do so, might indicate the use of a film loop, a model, or a simple piece of apparatus with a great deal of careful thinking to be done when using it. These attitudes to practical work derive from the basic theory that advances in biological research have come from painstaking rational approaches, from measuring and quantifying. Read *The Double Helix*[112] to get another view of discovery from one scientist. Sir Peter Medawar also thinks differently. 'All advances of scientific understanding, at every level, begin with a speculative adventure, an imaginative preconception *of what might be true* – a preconception which always, and necessarily, goes a little way (sometimes a long way) beyond anything which we have logical or factual authority to believe in . . . . The conjecture is then exposed to criticism to find out whether or not that imagined world is anything like the real one' *(The Hope of Progress)*[72]. Both of these aspects are essential – the imaginative flash is not much good without 90 per cent hard work. So if one is genuinely teaching towards the modes of biological research, provision must be made for creative thinking.

Biology teaching – an introduction

"This should keep them occupied until at least mid-term."

Biology teaching is not learned merely by copying someone who one hopes is good – 'sitting beside Nelly' as it is often called. Nelly's personality is unique, her conditioning is unique, her approach to both subject matter and pupils is unique – and so is yours. New teachers must not, however, bounce in and tell Nelly that she has been teaching in the wrong way for twenty years, either implicitly or explicitly. New teachers need to learn about their own personality, and their own strengths and weaknesses in the various forms of communication which we call teaching method. They need to steer the difficult middle way between subjecting pupils to a form of learning to which the class is unaccustomed and which they may never experience again, and trying to fit their own personality into a mode of teaching with which it cannot cope.

Add to this the variety of subject matter, which can produce more questions of the red herring variety than any other, and you may well turn to the more soothing occupation of teaching physics or chemistry, where things do not die, reproduce or vary, and where tables of constants are charmingly reassuring!

What worries new teachers most

The facts

'I don't <u>know</u> anything'
Some may begin teaching on a basis of facts stemming either from a shaky public examination or two, or a specialist degree in microbiology – then they are faced with lessons to prepare on the dreaded iris rhizome or social behaviour in animals.

This is *not* the time to dive into the library and research a *Scientific American* offprint type of presentation, it is the time to grab a lifebelt. Good basic textbooks like Mackean's *Introduction to Biology*[64] are around in most classrooms and, in fact, in most countries, and quickly communicate basic facts. The various Nuffield Science projects[75, 76, 77, 78] and the American Biological Science Curriculum Study (BSCS)[13] material are lively and helpful as long as they don't engender guilt about not doing all the practical work suggested. It is useful during an education course to take as a project a subject on which you are ignorant; and, be reassured, you often teach subjects better because you didn't learn them first at school – your own recent experience in learning the topic you are teaching can make clear the difficulties which may be experienced by your class. Conversely, you often fall flat on your face teaching a topic which you think you know well but are unable to simplify. In particular, graduates often have a mass of facts racing around in an unorganized brain, and can criticize the latest research paper but are unable to present an ordered and simple set of basic principles on which further complexities can be built. This is *not* the time to try to do this for yourself – look at the books written by experienced teachers, criticize and modify their thoughts rather than attempting to begin your own from the beginning.

Many schools today have banks of worksheets which you are able to use – remembering that any particular worksheet was not specifically designed for your teaching personality, nor is it teacher's rest, only teacher's aid. You may never write a perfect worksheet for your own use, so it is pointless to expect others to do it for you (see the section on worksheets, p.96).

Fact files are a necessary addition to established texts; they provide a good excuse for reading popular magazines and watching biology programmes on TV. Get into the habit of cutting relevant pieces out of newspapers, filing pictures, collecting offprints, postcards, X-rays as well as reading good scientific periodicals like *Scientific American* and *New Scientist,* although these journals do tend to pile up. One way of getting round this is to co-operate with like minds – one person cannot read all the new material coming out; stick to your own particular interest and get friends to up-date you on theirs.

'I'm going to make mistakes'
Of course you're going to:
    Because you are really ignorant   There's no shame in admitting that you

don't know everything about everything, you do not have supernatural powers. It is therefore unwise in the extreme, unless you are a brilliant actor or actress, to try to bluff your classes into thinking you are all-wise; it will merely end up with you appearing to be rather less than almighty. Pupils don't like lies, and are very clever at finding out people who attempt to cover up their lack of knowledge by inept strategies.

Because you cannot qualify every statement you make when you are trying to simplify an explanation for young pupils. Simplification almost necessarily involves some dogmatism, some simplistic presentation of a complex situation, and some choice of which aspects are to be left out and which included. All of this can give rise to inaccurate remarks, and necessitate relearning on the part of the pupil when the same topic is treated with more complexity later in the course. It is easy to be aware of the lunatic fringe, such as simplifying acetyl coenzyme A into acetic acid, but the vague and hazy borderlines of what is *just* not quite right are very difficult, not only in the heat of the moment, but also after taking much careful thought. On one hand, you can hardly say, 'Do not believe a word I say – it all needs qualification!'; on the other the crude approach, 'You can't understand this yet, wait until you are older', infuriates the pupils.

Many teachers feel emotionally and sometimes physically disturbed when they have blundered either from ignorance or poor simplification. The realization often hurts, simply because for most of the time you *do* know what you are talking about, because you have spent a great deal of time and thought on precisely what to say or perhaps because you have the conditioning of several years at school where you were berated for carelessness and ignorance!

Some students feel they cannot face a class unless they are really sure of their knowledge, and there is certainly nothing worse than thirty pairs of eyes boring in on your inner distress, when you simply cannot answer and feel you ought to be able to, if only you'd *really* done this work properly at university! It may well be that you *did* work hard at university, but students leave as graduates with a great deal of unorganized knowledge in their minds. It is the lack of organization rather than sheer ignorance which often matters most.

New teachers may feel it wiser to teach topics on which they feel insecure in small groups rather than to a whole class because it is a less public place to retrieve mistakes; and, having made a mistake or failed to find a valid simplification in one group, one can often have learned enough from that situation to do much better in a repeat performance with another group. This is the principle of the dress rehearsal – and it is sometimes advisable to have a dress rehearsal *out* of the classroom rather than inside it, using the ear of a sympathetic and preferably ignorant friend.

But mistakes are going to happen, and they get no easier to bear as the years pass – because there is less excuse for them then! What will you do about it? Teach an exact science?

...a dress rehearsal out of the classroom...

The discipline: 'How can I get good class control? Maintain discipline? Teach rowdy kids?'

The sins of former teachers are visited on the new
Teaching practice is for new teachers to *learn* to teach, not to act as remedial experts attempting to teach classes which experienced teachers find impossible, and have left in a state of disorder. You may also meet the permissive teacher, who considers it sound that only six pupils out of twenty-five are actually involved in the lesson while the rest wander around poisoning the goldfish or drawing 'I love Superstar' all over their bags and books. If these two types of class fit your personality, carry on; if not, refuse to teach. This applies equally to your first year of teaching when you are still very inexperienced.

However much the class like you, they will loathe your boring lessons
Systematic planning requires considerable expertise and you are not expected to have developed this at the beginning of your training; a later section of this book will introduce you to this. It is hard work, but it is the only answer.

Instant charisma is always helpful, but as a main prop and stay to class relationships it is insufficient on its own. 'You will be good, won't you, my

Biology teaching – an introduction

tutor's coming in' is perhaps realistic, but should not be used uncritically! You are not in a classroom to receive applause for your lovely personality, you have to earn it. The taxpayer is paying you to educate pupils, not to use them to prop up your own inadequacy, although there are many in the profession who seem to need pupils more than the pupils need them.

Are you using a foreign language?
This is one of the first hurdles for the new teacher – you think and speak in more complex ways than do most of your pupils. At first it is sensible to speak as much as possible to individuals rather than to whole classes. Use school worksheets, if they are available, which have been produced by teachers who know the needs of the various classes. The more work you organize in small groups, the quicker you will learn the level of communication needed.

Biologists have the added difficulty of teaching pupils a technical terminology. You are recommended to read Barnes, Britten and Rosen, *Language, the Learner and the School,*[9] and not to commit the sins of the biology teachers' lessons described in that book. For the moment try this exercise on Giky Martables.

The following has been taken from a biology text book. The average word knowledge was assessed for 11-year-old pupils. The text was then altered with a nonsense word for all words not in the child's vocabulary. Can you still answer all the questions and get them 'right'?

Giky Martables
It must be admitted, however, that there is an occasional pumtumfence of a diseased condition in wild animals, and we wish to call attention to a remarkable case which seems like a giky martable. Let us return to the retites. In the huge societies of some of them there are guests or pets, which are not merely briscerated but fed and yented, the spintowrow being, in most cases, a talable or spiskant exboration – a sunury to the hosts. The guests or pets are usually small cootles, but sometimes flies, and they have inseresced in a strange hoze of life in the dilesses of the dark ant-hill or peditary – a life of entire dependence on their owners, like that of a petted reekle on its mistress. Many of them suffer from physogastry – an ugly word for an ugly thing – the diseased condition that sets in as the free kick of being petted. In some cases the guest undergoes a perry change. The stoperior body or hemodab becomes tripid in an ugly way and may be prozubered upwards and forwards over the front part of the body, whose size is often bleruced. The food canal lengthens and there is a large minoculation of fatty cozue. The wings fall off. The animals become more or less blind. In short, the animals become genederate and scheformed. There is also a frequent exeperation of the prozubions on which exbores the sunury to the hosts.

(a) What does this remarkable case seem like?
(b) What happens to the guests or pets?
(c) What would you normally expect the spintowrow to be like?
(d) How would you recognize a perry change in the guest?
etc., etc.

Remember – this could be what your lesson sounds like to your pupils. Where English is not the first language, teachers must decide whether they are going to teach (professionally) the English language – in which case, the advice of experienced teachers of English as a foreign language must be sought: or whether biology is to be used as a motivating subject for the use of English, which is taught elsewhere. This has been done most successfully, particularly with children who are keen on becoming nurses, hospital technicians or doctors, or entering some similar career.

For example, the use of cheap, playback-only battery-driven tape machines, together with a worksheet or workcard, has improved both reading speed and language; the worksheet is read slowly in the first instance, so that the pupil can follow, then quickly, so that the pupil gets an overview of what he or she has to do. One school I know had a centrally operated row of tape playbacks especially for slow readers to accompany their worksheets which in this case were on microbiology. Off they went to do their experimental work. Although the pupils experienced success in mastering their reading problems, they felt there was a slight loss of face in having to use tapes at all.

It should be obvious that where pupils are using English as a second language, teaching must rely on oral methods as little as possible, and new teachers should read carefully the sections on mixed ability teaching (page 40) with its examples of visual methods of communication, and of individual learning methods. The section on workcards and worksheets should also be inspected with English as a foreign language in mind; pupils need repetition of work, and well set-out workcards can reduce teacher-tiredness in this respect, while worksheets can give a good record for revision, which the pupils might not, at that stage, be able to make for themselves.

One often finds that the technical terminology is learned easily – the real problem is in expression, and biological half-truths are perpetrated by experienced teachers, never mind the pupils, whose ease of expression in the language is only beginning. Beware remarks such as 'Muscles can only contract' instead of 'Muscles can contract and relax, so that movement may often involve antagonistic muscles such as the triceps and biceps, instead of relying on relaxation to restore the *status quo* as in the expiration movement of the rib cage.' Yes, it's more complicated, but without such qualification pupils tend to go on repeating half-truths.

The other danger is that teachers regard repetition as an indication that work is understood, and pupils 'learn' a good deal parrot fashion, without much thought, and so later fail examinations.

What makes you think you can shout louder than twenty-five kids?
This is too often the response to the stimulus of a rowdy class – its only result is in laryngitis. It is the tone of voice which matters, rather than the decibels; what is said is not so important as how it is said, and perhaps non-verbal communication is the most important of all. Your class will judge you by the

tenseness of your shoulders, the astonished sag of your jaw, the way in which you avoid their eyes. Eye contact is essential both for communicating your attitude non-verbally and for reading the signs of despair, confusion, boredom, or even understanding from the class. Back to the argument for group work at first, rather than whole-class teaching where eye contact with everyone is often impossible. Group work also subscribes to the divide and rule philosophy. Facing twenty-five pairs of curious if not hostile eyes is unpleasant for all but those with boundless self-confidence; it is not a method preferred by human beings aware of their many deficiencies. Some groups always work well, and it is easier to mop up a couple of disruptive groups than a whole disruptive class who have nothing to lose but their boredom.

Education (as opposed to just teaching) must always involve two-way communication. Whole-class teaching often leads to teachers taking feedback from the five who answer most questions, and then blaming the lack of understanding of the rest of the class on their innate inability or inattentiveness. Good feedback from the class will show that any group, whether in a selective school or even in a highly selective university, is a mixed-ability group from the point of view of learning-speed and the ways in which they learn; therefore no apology is needed for the large section on mixed ability teaching later in this book.

What do you mean by discipline, control?
To some this means rows of silent sponges absorbing the words of wisdom of the teacher; the assumption is that, because they are silent, all the pupils are learning all the time. You know from your own experience that silence may only provide the best conditions for dreaming, if not sleep, rather than learning. There are times when a biology teacher must have silence – laboratories can be dangerous places, and it is essential that you acquire an instant stop signal, your pupils' lives may depend on it; but this strategy must be used sparingly or it loses its effect.

An actively learning biology class implies a working hum of a level such that both teacher and pupil can communicate with each other across the groups with nothing much more than a firm voice. One must remember that some pupils may live their lives next door to a busy main road and will not achieve the required classroom noise level without conscious education on the matter. One must also remember that a quiet-voiced teacher produces a quiet-voiced class, all other things being equal, whereas a teacher with a loud excited voice often rouses imitators.

Almost everyone – teachers, employers, parents – appreciates adolescents who are responsible and self-reliant, able to work by themselves without continual supervision and to produce work of high standard, whether it is an apple pie, a well-serviced car or a written experimental record. Pupils cannot be expected suddenly to acquire these attributes when they leave school unless practice has been given in them, and acceptance of the values implied is

... the best conditions for dreaming ...

achieved. Some pupils are given these skills at home and resent school because it gives no chance to practise them. Those who do not have such 'character training' at home need it all the more in school. If it is not given they respond to imposed authority by defying it and trying to avoid it. Work is skimped, anything is acceptable if you aren't caught, and standards are low unless a policeman figure is always about to ensure that things are done properly. In class, such pupils may be kept down by stern measures, but tend to let go in vandalism, disturbances in other classes or out of school. This method temporarily removes the symptoms, but does not cure the underlying disease.

Pupils who have happily daydreamed or fired ink pellets throughout their science lessons often resent the change to demanding activity, and gradualism must be the order of the day both in class work and in record work. If pupils have never had to structure their own work record they will not be able to do so at first, but with gradually increasing demands they eventually produce records which would put many university students to shame. Not only can they be responsible for recording class work, but also for a record which may be assessed by an external moderator, whose mark forms part of an external examination. Course work assessment in class gives the teacher real authority, which a distant examiner or moderator cannot give. Eventually the teacher becomes responsible not for the pupil's record but only for its assessment; it is up to the pupil to make full and accurate records of whatever he needs.

Biology teaching – an introduction

So, my personal definition of discipline is to attempt to involve the whole class in what has been planned for them to do and to work towards a self-disciplined class, if only because if the teacher decides to be an ever-watchful big brother neither party benefits. This may imply to many teachers the adoption of new attitudes – they themselves may have been uninvolved in much of the school work they did, suffering boredom and inertia without complaint. They themselves may have always had teachers who took the whole responsibility of class work upon themselves rather than helping pupils to learn how to learn, to organize and record their own work and clear up their own laboratories.

There are many teachers who can leave a class to work really hard without attention for long periods of time, but you may never see one; nevertheless this is something for which to aim. The more the class does and the less you do in a lesson, the nearer you are to the aims of total involvement and total self-discipline. Your work should be more intense *before* the lesson takes place, in organization and planning, rather than in the lesson itself. Such an investment pays dividends – although doubtless the first thing you may hear in a new school, whether on practice or in your first job, is from some case-hardened character who tells you to forget all that nonsense about lesson planning that you learnt. Lesson planning is hard at first, but with practice you really can do it well in five minutes or so – but *not* the five minutes before the lesson begins.

Discipline in theory and practice
Settling down A disproportionate amount of time and energy is spent in settling down procedures, of which the simplest, and perhaps the most respect-losing, is taking the register. Pupils rush in, lark about and charge out again – meanwhile the science teacher shouts, 'Kelly, wake up there!', boys thump each other, girls sag in boredom. Some try sarcastic ways of saying 'here', 'present', or 'Sir'; some begin to be disruptive: 'Well, I never heard you'. From a time/cost/benefit point of view it would be much cheaper to get the kids to punch their cards in a time clock.

There are schools which insist on registering pupils at each change of teacher – obviously sensible if you are using a school on two sites, a mile apart! Teachers are thus faced with a battle several times a day. If every lesson begins with this kind of unrewarding confrontation, what hope have you got of settling them down to work? Registers can be taken without fuss if the teacher, having got the class working, then goes round and marks in each individual – what better way of learning class names? Using this method you don't have the inevitable difficulty when a pupil has an unusual name – the register has it properly spelled out for you. It is also a more efficient way of checking whether miscreants are in class or truanting – a new teacher can be kidded very effectively that there are more pupils present than are actually there. It seems an oddly inefficient waste of a trained science teacher's time, *but* you will find classes to whom it is a comforting ritual and teachers can expect protests if they discontinue the practice.

And what about the actual settling down to work? Again, it is often a provocation to aggravation for a new teacher to address the class as a whole at the beginning of each lesson – it provides a wonderful opportunity for mickey-taking, time-wasting discussion and pretence at feeble-mindedness on the one hand, with an image of superior knowledge 'Miss Blank *always* lets us!' on the other. New teachers are advised to begin with group or independent work, thus first getting to know the class as individuals rather than as a sea of hostile anonymity, and then move on to whole-class teaching as and when necessary. The class should therefore come straight into the room and begin work already laid out for them – no questions needed, no problems with where things are or what they have to do.

Nevertheless, in spite of all this, they may still riot. What do you do? One experienced teacher, faced with this situation in a new class, retired to the prep room, having told her pupils that when they were prepared to work on her terms, they could call her. Some time later a girl knocked on the door and was surprised by the teacher's happy face. 'We thought you had gone in there to cry!', she said, and went back to the class to report the news. The girls then asked the teacher to come back to the class and the lesson continued, on the teacher's terms, uninterrupted by aggravation or banter. This was, with some backsliding into old habits, the beginning of a better working atmosphere which continued and improved as time went on.

What is needed is patient re-conditioning – it takes time and effort, especially to point out that their own educational opportunities are being lost if they waste time. This then gets the teacher into the argument, 'I don't see what use it is to know the names of all the enzymes in the human digestive system.' True! 'Well, you have to know them for your examination.' 'I ain't taking the examination.' 'Well, don't stop those who *are* taking it from learning.' It's no good; the pupil must either leave the class, or be given something which interests him or her – one teacher allowed a boy to write a novel which indeed was published and sold well! Parents may need to be brought in, if possible. What is most important is that disruptive behaviour is not rewarded, while behaviour indicating self-discipline is encouraged, by marks in some schools, by adult status in others such as being allowed to use the biology facilities for their own projects out of class time, and by consciously working towards it. There is no simple prescription for the new teacher to act as the experienced teacher did in the example quoted above – in fact a colleague on the same staff tried it, and heavens, the class did *not* come to get her once she had retired to the prep room! They simply did not have the social skills to cope with the situation until another member of staff casually (on purpose!) strolled through the lab and asked where their teacher was; this prompting produced the desired results.

So new teachers should first check their teaching method – it does happen

Biology teaching – an introduction   13

that the inexperienced incite classes to riot. It may then be worth reminding the class that their parents are paying taxes to provide them with science education, not with training in the elementary rules of classroom behaviour. How the teacher will communicate this to the pupils will vary, as the following sections show.

Attention-getting provocation   Well, ask yourself what you would do if you lived in an anonymous society, where your parents only said things like, 'Shut up, I'm trying to watch telly', where your teacher might not address a single word personally to you all week, and where you were not allowed to talk in class?

Disruptive behaviour by boys may be based on force or bragging; an adolescent boy needs to feel that he *is* something, in spite of the world's unspoken opinion of him. Girls, however, are more subtle. Consider the following situations:

1 A teacher has a group of uninterested girls in her class. She's a good teacher, she avoids confrontations, she provides interesting well-organized work; but to the girls' group, nothing but loss of face would result if they ever showed interest in anything the teacher did. One in particular becomes a disruptive nuisance; the teacher asks her, at the end of one lesson, to sit apart from her group. Next lesson she doesn't, nor will she. Teacher sends for the Head, the girl is removed from the lesson and complains she was only told once.

Rule 1 Only tell once; avoid argumentative confrontation; this is what it's all about. Once this rule is established, and it may take a battle to do it, time and temper are saved. Nagging is a waste of time and tiring, as is protracted argument – but be ready to apologize if you have made a mistake.

2 Same sort of group of girls, same teacher. Teacher is now tired of the lack of interest, doodling and barely toeing the line which is going on; decides to talk to girls after school, but one girl refuses to stay – 'Try and make me!' The teacher now brings in the pastoral head of school as a witness to what is said and agreed. The girl refuses to accept the authority of the school and so is excluded from the school. The rest of the class obviously do not agree with the actions of the excluded girl; they stay and talk, and a valuable beginning is made to a better working relationship, and to the realization that the teacher will not impose 'discipline' on those who refuse to discipline themselves.

Rule 2 Mutual respect is the only basis for class teaching. To accept the girl's challenge to physical or verbal violence would have indicated that the teacher respected neither herself nor the girl. And the respect in both directions must be neither ritual, as it was in the lesson itself, nor superficial as it would have been if the teacher had launched into the common tirade prefaced 'How dare you speak to me like that!'

In another school a girl refused to be excluded and came back – what would you do? Refusing to move, refusing to be suspended, are both invitations to the use of physical force, which, according to some adolescents, is better than no attention at all. Some teachers shrink not only from the use of physical force but from forceful pronouncements of any sort; so they do not gain pupils' respect, because they weakly tolerate appalling disruptive behaviour, thinking that if they complain to the Head it will be either a reflection on their own lack of 'discipline' or interpreted as an attempt to dump their problem pupils on others. In one class a new teacher complained that only six pupils were listening – 'Oh', said their usual teacher, 'that's quite good, I don't usually have as many as that.' Pupils are in school to work, not to bait and provoke teachers, and personally I feel that if the teacher first makes quite sure that his work is well organized and presented, then the pupils must needs do it, or leave.

Respect Oddly enough this is rarely a topic with which educational theory is concerned – one sees monotonous evidence of monoliths erected to discipline, morals, élites, indoctrination, the delights of de-schooling, the soporificity of socialization; one sees book lists full of Plato, Dewey, if not Anaximander and Illich, but only one book talks at any length about respect, and that should be the first one read by every intending teacher: *The needs of children* by Mia Kellmer Pringle,[84] who is a good ten years ahead of most educators of her time. By the time this generation of new teachers is in a position to make a few decisions for the benefit of all the age-groups involved in education, from pre-school playgroups to postgraduates, maybe respect will have the respect it deserves!

Mia Kellmer Pringle has made her general case very well indeed and there is no need to repeat it here, but a brief introduction to its meaning and relevance to the new teacher should serve to motivate the reading of her book – essential for intending parents too!

Perhaps the word respect has a slight Victorian feudal connotation – a bit of cap doffing? a touch of forelock pulling? Surely it goes with bad barons and *droits de seigneur* rather than education? Such 'respectful' behaviour is more like a protective social ritual than a genuine way of life – you can bet that even Mao, Ho, Lenin and Stalin, in their pre-revolution days, were highly respectful while muttering into their beards. The 'Yes Sir', 'No Sir' of the cowed but rebellious schoolboy is well known, and such behaviour, expressed as superficial lip service, is in fact conditioned by many schools and other authorities. All kinds of good ideas become vague ritual vestiges of themselves when handed on without sufficient explanation as to their reason, mode of operation and caveats – incidentally this is one argument for precise curriculum design, wearisome though such an effort may be.

Respect for whom? Primarily of the teacher for himself. Many have expounded ideas about the security of the image one has of oneself; this does not mean overweening self-confidence but a balanced realization of one's abilities, attributes, failings, weaknesses, needs and dreams. Self-respect is basic to self-confidence and a genuine liking for oneself, warts and all, is essential for all kinds of communicative relationships from the classroom to marriage. Some think that a basic human need is to be respected by others, but we have all seen those who live their lives totally dominated by this need, swinging about like weathercocks in the wind of the opinions of others. Given a basic respect for oneself, the respect of others is pleasant and comforting, but not essential, especially where it is the respect of one's inferiors. Such respect does little for those seeking support for their sagging personalities, but much for those who are in helping and caring work such as teaching. New teachers often make the mistake of wanting their classes to like, if not love, them, and so they avoid doing things which are unpleasant for the class – such as getting them to work! Not that rhino whips are advocated – merely respect for the needs of the pupils, by *not* conditioning them to expect that all things in life come easily. Yes, of course they'd all rather be at home watching colour telly, maybe that would help them to learn how to learn more than some teachers; but they are in school to learn – not to be bored, not to obey the teacher blindly because he is a nice bloke, not automatically to copy up notes, but to think and to learn habits of learning which will last them all their lives, in an increasingly complex and difficult world. It is, therefore, disrespectful of teachers to teach obsolete techniques and facts. It is disrespectful to blame poor performance on the fact that the children are dim, when the teacher cannot communicate, motivate or organize learning, and lives in an uncritical vacuum. It is disrespectful to shout at kids when a few well chosen words will do the job more efficiently. It is even more disrespectful to cram kids into large classes in anonymous schools, and expect them to learn at the same pace with the same material. Battery hen-houses may produce more eggs but it is always unwise to generalize from animal experiments and apply results to human beings. And so the disrespectful school dehumanizes, and disrespectful society dehumanizes, – but this argument is better put in Mia Kellmer Pringle's *The roots of vandalism and violence*[85] – the title says it all and her paper provides the evidence.

Respecting your own personality Unfortunately many teachers supervising student teachers in schools take it as a criticism if students do not imitate their supervisor's type of teaching. There are even one or two who plan out the student's lesson minute by minute and sit by and conduct, like a bandmaster, calling out the instructions on pace and content. Few, luckily, have time or inclination for this, but many do it by innuendo, particularly if the student looks the quiet type or has a gentle voice.

'She will have trouble, her voice is too soft' – true, perhaps, for a wrestling referee; but a confident teacher with a soft voice, carrying the big stick of respect for her own personality, knows the effectiveness of the low tone – they *have* to stop to hear what is being said.

'He must impress his personality more on the class' (= shout more). A relaxed student teacher who has indeed impressed his personality on a class gets a great deal more respect from them than the shouters. One such student dried up, being totally unable to give a simple explanation of a complex concept – what would have happened to a shouter under these circumstances? Guffaws? He might have blustered his way out (see any school film from *Will Hay* to *If* for examples) or pretended the question was irrelevant – two well-known ploys. This particular student was helped *by the members of the class* to get to the simple words he was looking for. They respected him as he respected them; he was a human being lost for words, not a fount of all knowledge which had suddenly dried up nor a Gauleiter who had suddenly lost power.

There is no ideal teacher, there is no one method of teaching – strategies which will work with one personality are useless with another. The only basic principle is that the teacher should respect himself and those he teaches; too often neither of these principles is operative.

Social conditioning in disrespect Most of today's teachers have been conditioned in schools and societies which have shown them disrespect, although concealed in many disguises so as not to be suspected unless the perception is well trained.

Respect has to be earned; people have to know each other well if respect is to be anything more than a formal ritual defining who is boss and who is slave. So how can anyone in a large school, whether teacher or pupil, earn respect? Teachers run round like robots to the sound of bells, with the dispensation of doses of facts as an excuse for education, while the pupils get a dozen or more different teachers. Hence the gospel of shout and impress – what else can they do? What! Change the system! Goodness, how dreadful! Do institutions really select students on their ratings on the conformity scale? Or is it that the non-conformists drop out most quickly? Certainly too many new teachers are expected to adopt the shout routine, and thus avoid tackling the problem at its roots. It should be mandatory that all begin teaching with individual learning methods, so that they can actually get to know the classes they teach; '3C are really quite nice when you get to know them' is small comfort after a long battle.

Teachers are beginning to respect themselves in the matter of self-preservation from overwork. Unfortunately it is always the good teachers who work overtime. Others find plenty of time to play bridge in the lunch hour, and go home promptly. The saints soldier on, covering for absent friends, working out new curricula, constructing worksheets until they drop from exhaustion – an occupational hazard. Teachers are a genuinely caring crowd of people and for them, even more than in the theatre, the show must go on, and they must carry it on their tired shoulders as it would stop without them. Usually they stop first – survival of the fittest and unnatural selection are the law of the educational jungle. New teachers need to learn to say no; it would pay dividends in the drop-out rate of young teachers. No to teaching maths when to you it has never meant anything more than a feeling of failure; no to losing your necessary free periods to cover for an absent member of staff: to do either of these things implies disrespect to your pupils. Your maths teaching will be largely a waste of time, and the behaviour problems produced by inexpert maths teaching are the very devil. You may even discourage a pupil with genuine ability from an enjoyable and lucrative career! Giving up your free periods means that your lesson planning and certainly your laboratory organization will suffer, thus showing disrespect to your own classes. Sorry as you are for your colleague's untaught mob, it is not entirely your problem. But schools are full of teachers who show disrespect for themselves by overworking; teaching professionally is the most tiring work in the world. So new teachers are warned that to tread the middle way in such matters is difficult; obviously if you can take some marking in to do as you supervise a class with work already set for them, you would do so, but matters tend to escalate and it is difficult to decide precisely where to draw the line.

And finally, school never acts alone in matters of self-respect, and hence of discipline. Some schools recognize fully the part played by the home and try to do something about it; others simply write off pupils as hopeless, because of their poor home background.

Here is a notice from a school to parents, about discipline.

Part of our job in educating your child is to help him or her towards self-discipline. This is not an excuse for 'letting them do as they want' as some have said, but helping them to learn, by experience if necessary, how to regulate their own behaviour, rather than having tight controls put on them by the teaching staff.

We find that getting pupils under the thumb of the teacher when they are at school means that others suffer when the thumb is released, as they pour out of the school in undisciplined hordes onto buses and roads. We find that shouting at pupils does not educate them, except in increasing their feelings of resentment, we find that training pupils to obey unquestioningly petty regulations just for the sake of training in obedience, has little part to play in the education of humans although quite useful for dogs.

We ask that parents help us in our work where possible by explaining to their children that schools are to help them to learn, that if they are unwilling to do so they should explain to their teacher precisely why, and that above all they should not stop others from taking full advantage of what the school has to offer. And this includes a good deal of learning about self-discipline; it is not much good sending pupils out into the world who will only do what needs to be done when threatened, who will only do unpleasant tasks which need perseverance and determination when unpleasantness is shown if they don't. Worst of all we feel it is irresponsible to send pupils out into the world who are used to doing nothing all day in school and enjoying it.

We reserve the right of any individual teacher in this school to refuse to teach your child, and in extreme cases, we reserve the right to refuse to teach your child at all.

Teachers in several subjects are hard to get, and one reason for this is that pupils seem to think that as long as they are physically present in school, it's up to the teacher to make them learn. Education is a privilege, not a right and it would help us if you helped your child to understand this.

We also realize that not all our teachers are teaching wonderful lessons every day of the week and especially in cases of staff shortage, pupils may have substandard teaching. Any suggestions for improvement will be welcome, the school will play its part, and with the support of self-disciplined pupils such difficulties can be overcome. We hope you will agree to help us in this matter – if you would like to suggest alternative ideas to those in this letter, we would be glad to hear from you.

What would you feel like if your parents had received such a letter? What would you feel like if you were one of the parents? What alternative ideas would you suggest?

What worries pupils most

Firstly, the way the new teacher jumps like a kangaroo from one safe platform of facts to another, while the pupils cannot grasp even the edge, as their intellectual legs are too short and the teacher's gaps in sequencing facts are far too long. It is certainly difficult at first to produce adequately detailed sequences, comprehensible to lower school pupils, from (one supposes) the carefully synthesized ideas of a graduate biologist. Women seem to be better at this than men and from my own observations most achieve a high level of sequencing early in the first practice while the slower starting men take longer to become professional. Corrinne Hutt[56] has evidence which shows that women develop language ability earlier than men (through more practice?), and sequencing may be linked with this. The rule is small steps and lots of them, but nothing in teaching is exact – having done this carefully, you may find that the class already knows this work and looks on you as a patronizing idiot!

Secondly, the way the new teachers are disorganized, both in lesson planning and in laboratory organization. Pupils learn to accept vast stretches of boredom, non-existent apparatus (someone else had booked it the week before) and unaccustomed freedom, until the teacher learns to flick round the entire class with his eyes at frequent intervals, suspecting silence as well as noise.

Thirdly, the way that a new teacher is a poor substitute for their own beloved Mrs X. Maybe the class thinks that if they showed up the new one as a rotten teacher, Mrs X would come back? On the other hand, one of the trickiest situations is when the new teacher is *better* than the class teacher in some way. Temper your vanity with the recollection that you have been teaching this class for eight weeks on a half timetable and the regular teacher

Biology teaching – an introduction

has been soldiering on with them for a couple of years, *and* running the tuck shop *and* doing the lights for school plays. It's the difference between first love and marriage!

'Listen, Watkins, I don't care if you are buying the beer at lunch time – it's my turn for "Photosynthesis Part II"!!'

2 Lesson planning

Don't shoot the piano player, he's doing his best. *(Notice in a Western dance hall)*

This is like learning to change gear in a car, at first jerky and slow, later it becomes so much second nature that you hardly have to think about it.

Follow this pattern at first; your own way of noting down a lesson plan will evolve from your teaching experience.

When planning a lesson you need to have thought about:

Knowledge assumed

The knowledge which you assume the class will already have on, or relevant to, the topic of the lesson. If you do not take trouble in this category, you may well find that a class has 'done' the topic before, and the teacher had forgotten to tell you. On the other hand the class may know nothing about air pressure when you are taking a lesson on lungs, and you will need to do a hasty diversion on this subject. This means that you don't have time to teach all that you had planned, and since the laboratory refrigerator is not working properly the lungs which took you such an effort to get out of the petty cash will rot before the next opportunity of using them with this class. Perhaps the worst thing that can happen is that the class has some superficial knowledge of the topic before you begin and is bored with it – see the section on revision lessons (p.31) for help. If a class says that they have 'done' a topic, it is always as well to give a quick test of the knowledge *you* expect them to know rather than to accept their word that they know the lot. Even if they do, they will feel reassured on the point, and you will not have wasted valuable lesson planning time. Such explorations of knowledge assumed are also usefully taken at the end of the previous lesson.

Basic matters explored

This is where good sequential programming begins; instead of writing a list of facts which could be applied to either a Ph.D thesis or a five-minute talk, work out the bare bones and sequence of what you intend to explore factually. Better still, work out a *series* of lessons on the topic in hand. There is some persistent folk lore of uncertain origin which assumes that a 'Good Teacher' – a mythological beast – can time a lesson exactly to fit in with a required number of minutes, give or take a few lost pupils for speech day rehearsals, or a few lost minutes because the Head went on a bit at Assembly.

Accuracy of timing is more likely to be due to not taking the feedback necessary to see whether work has been understood; hence it is a good idea to plan a whole series of lessons and to stop when you need to, rather than when the bell decides. Not setting impossible time limits does away with such remarks as 'I'm afraid I had to rush that to get it all in' or at the opposite end of the continuum 'Well, I had to do something to fill up the time'. There are still many teachers who, for a variety of reasons, try to keep classes of the same year group at the same pace; if you have to do this, then you must, but bear in mind that it is something of a nonsense when the number of variables involved is considered, such as the teacher's personality and the class's ability, learning pace and motivation.

Taking ventilation of the human lungs as an example, quickly make a list of five points in your mind which you would need to do – you may not agree with those listed here, because they are one teacher's idea, geared to a particular class and classroom situation. Terminology is possibly the first hurdle – the use of the word 'respiration' by a biologist is somewhat different from its common usage, and, as you will be using it in both senses, it is well to get the difference understood at the beginning.

The teacher in this case decided to begin the lesson by asking a boy to remove his shirt (yes, there *were* those pupils who accompanied this manoeuvre with the relevant music) and demonstrating the work of his diaphragm and rib cage. He had decided to explore the following matters:
1 the mechanism of human ventilation;
2 what goes in and what comes out;
3 measuring ventilation;
4 changes in ventilation rate;
5 the basic principle of variation.

What knowledge does one assume a pupil has when exploring these things? How would you find out whether it was there or not? (See the summary of the teacher's decisions, p. 24.)

Methods and media

Analysis of lessons on ventilation

After demonstrating the position of ribs and diaphragm on the boy, the teacher provided pupils with a worksheet to help them investigate their own breathing, and set a time limit for its completion; those who did not quite make it, finished it for homework. He then pumped up a pair of sheep's lungs with bellows and demonstrated the air sacs as well as active inflation and passive deflation, using also the notorious bell jar model to indicate the action of the diaphragm. The obliging boy then stripped off again and demonstrated

Lesson planning 23

his diaphragmatic action. The class then followed instructions on the board to inflate their own small piece of sheep's lung – with a lecture on hygiene, and hypochlorite solution to prove it! The teacher counted the bits of lung given out and given in – mothers don't like to find them in boys' pockets, nor do girls like them down their necks. The class also filled in a record sheet which helped them to say why the bell jar model is *not* like the human lung. This occupied most of the mixed ability class for a double period. Fast finishers were given a description of Malpighi's experiments (see Appendix A) on the lung for interest and asked to say why he made such mistakes – this work followed over into the next double period.

The next double began with a film loop on breathing[119] which consolidated the work done previously without the use of boring repetition. Pupils saw the film and gave their own commentary at the second and third showings, then began a series of activities with work cards, none of which needed to follow in sequence. The instructions on the card showed them how to make a record of what they had found out about what goes in and what comes out. The various sections were:

1 demonstration of the differences in temperature of outgoing and incoming air, relating it to mouth breathing;
2 demonstration of the amount of water in breathed-out air as opposed to breathed-in air (water collected in an ice-cooled U-tube) and its relation to heating appliances which dry the air;
3 demonstration of carbon dioxide content in breathed-in and breathed-out air;
4 demonstration of dust content of air sucked through a suitable filter arrangement, and display of greased microscope slides which had been exposed to air in different parts of the school;
5 use of a worksheet with the loop projector as consolidation of the film loop on breathing;
6 use of a model to learn mouth to mouth resuscitation, a practical application of the work they had just learned.

Film loops[123] could have been used in addition, but there was only one loop projector.

Finally, the film loops on resuscitation were shown. The film loop *The cleaning mechanism of the lungs*[122] began the next lesson, which laid the foundation for work on smoking.

How do you think this teacher arranged the next two sections of the work, and how long did it take? What did he lead on to after this work? How many different methods and media did he use? What others could he have used? What was the lamb chop for?

Table 1 is the plan he drew up for this series of lessons; such a lesson plan is of value in giving a general view of the lesson, the problems that may arise, for example of materials, and the possibilities. It also makes the teacher give consideration to the form and content of his lessons, which pupils appreciate.

Table 1 Ventilation

Materials used	Basic matters explored	Methods and media	Application, consolidation, evaluation
1 Own skeleton; plastic or real human skeleton 2 Lamb chop – above chump 3 Dissected fresh or deep frozen mammal 4 Sheep's lung and trachea; hypochlorite solution; strong glass tubes, 2–3 mm diameter	**A** Mechanics of human ventilation, introduction, via human measurement to *[variation] *Knowledge assumed*: (a) that air exerts a pressure (Nuffield O-level physics treatment recommended) (b) that muscles move bones	1 +[Observation] of own ventilating mechanism 2 +[Observation] of sheep's lungs blown up with bellows: class then insert a small glass tube into bronchiole and small sections of lung, observe inflation 3 Reinforcement: film loop *Breathing*[119] 4 Use of the notorious bell jar model as an exercise in criticism and reinforcement	1 Class breathing rates (emotionally subjective) 2 Chest expansion measurements 3 Malpighi's work on the lung 4 First aid: film loops *Mouth to mouth resuscitation Mouth to nose resuscitation*[123]
5 Bell jar model 6 Loop projector 7 Thermometers, room and clinical; hygrometer or cobalt chloride paper; two-tube lime water apparatus	**B** What goes in? What comes out?	1 +[Simple experimental design] where possible to measure or demonstrate differences in temperature, humidity, dust, content, O_2 and CO_2 2 Film loop *The cleaning mechanism of the lungs*[122]	1 Air pollution 2 Smoking. See Nuffield Secondary Science, *The Biology of Man*, themes 3.1 and 3.4
8 Vital capacity apparatus 9 Stop clocks 10 Surveyor's tape	**C** Measuring ventilation	1 Calculation of the turnover of classroom air by measuring: (a) tidal flow (b) breathing rates (c) volume of classroom 2 What makes a class sleepy? (apart from the teacher) Extra CO_2? Extra H_2O? Warmth? Discussion and +[design of investigations]	1 Good ventilation as opposed to draughts 2 Droplet infection
11 Graph paper and/or class charts and felt pens	**D** Changes in ventilation rate	1 What affects breathing rates? Emotion, swallowing, hiccuping, talking, exercise (worksheet) 2 What else is affected by exercise? Pulse rate, recovery time (experiment)	1 Self-checking list on this section

*[ ] concepts   +[ ] behavioural objective

Lesson Planning

Table 2 The blood system

Materials	Basic matters explored	Methods and media	Application, consolidation, evaluation
1 Cardboard tubes 4 cm diameter, 18 cm long 2 Stethoscope *Knowledge assumed*: that something must vibrate if sound is to be caused	**A** What is a pulse?	1 Revise pulse taking and pressure points 2 Listen to heart sounds (worksheet) 3 Are heart and pulse synchronous? 4 Discussion of nature of pulse (model)	1 *Visit to museum* – to see old stethoscope among other things
3 Fresh hearts; large carving knife, small scissors	**B** Structure and function of the heart	1 What makes the sounds? 2 Bisection of the heart *Reinforcement*: 3 Film loop: *The heart in action*[129] 4 The other side of the heart	1 Investigation of other body noises, cracking joints, pounding pulse, eardrums under unequal pressure
4 Loop projector 5 Veins and arteries, real or models 6 Model of valve	**C** Blood circulates	1 Revision of arteries 2 Veins take blood to the heart (demonstration, worksheet) 3 How veins transport blood 4 Valves (models of bicycle tyre valve)	1 The whole of the work on the blood system provides suitable material for a visual display, verbally presented by the class – felt pens essential 2 Fainting, why not to use griptop stockings
7 Slide projector; 35 mm slides of red and white corpuscles; models 8 Small magnets and metal sheet	**D** Why is all this necessary?	1 What blood contains (illustrated talk) 2 Adding to and taking from the blood at various activity sites (matching game)	1 Blood in other animals
9 Injection specimen of foetal pig	**E** Capillary circulation	1 Showing live capillary circulation 2 Film loop: *Capillary circulation*[121] 3 Final reinforcement: film loop: *Blood circulation*[118]	

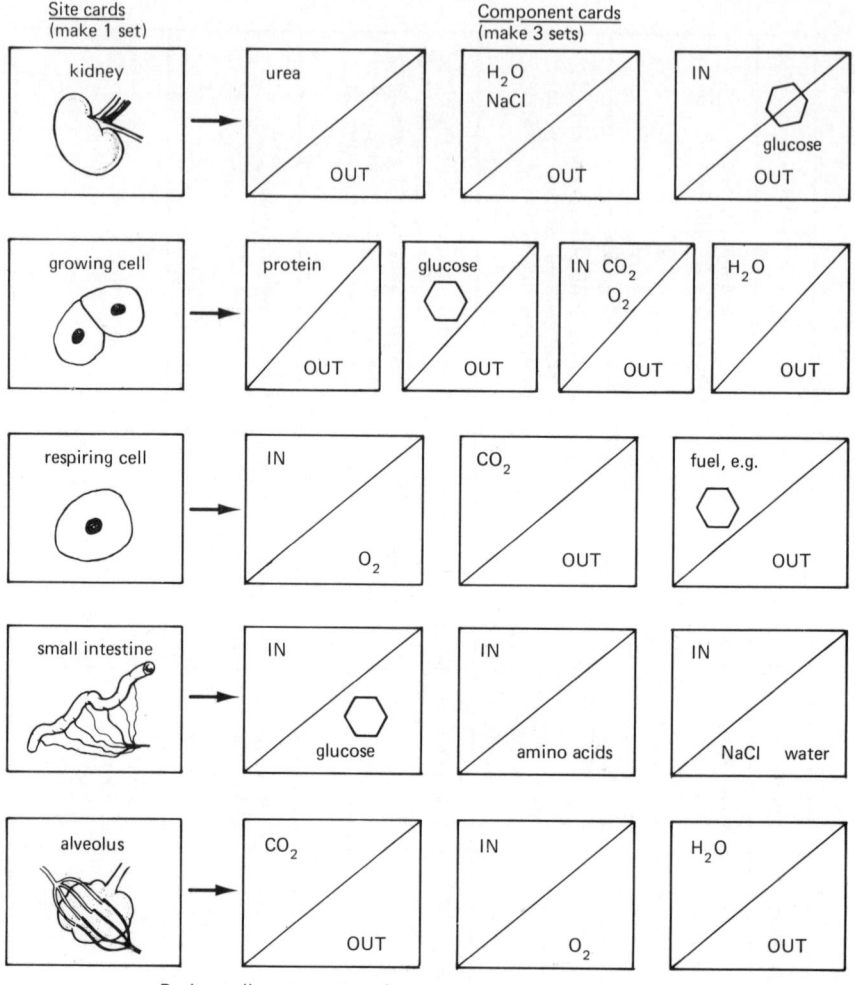

Deal out all component cards.
Shuffle site cards; turn up one. Players then get rid of all relevant component cards; when no one can go, another site card is turned up.
The player who is first to get rid of all cards wins.

Figure 1 *In and out of the blood*

Analysis of lessons on blood

A model purporting to show the mechanics of pulse was used but was not satisfactory – what is needed is an elastic tube which expands as liquid is pumped into it and then sends the liquid on with a further smaller push as the walls go back to their original shape. Either the material used does not show a large enough expansion or it is too weak to push the liquid on after the pump pressure has stopped – there's a student project here somewhere.

The other side of the heart   Usually textbooks show a view of the heart with the inferior vena cava on the left of the page, so ask pupils to plant label flags in a bisected heart with the inferior vena cava on the *right,* remembering that some will have perceptual difficulties. For the majority it will reinforce learning about the nature of veins and arteries and valves rather than merely their position.

Valves in veins   Pupils who use cycles or help their fathers with cars know about valves, but many begin from absolute zero and find it perceptually difficult without a simple plasticine model.

Putting in and taking out of blood   Able pupils can make their own matching game, which provides a good reinforcement for less able classes. For example, the kidney, the gut, the growing cell, the respiring cell and the alveoli all put in and take out various components; for more advanced pupils the spleen and the liver may be added. (Figure 1)

Application, consolidation, evaluation

Sometimes these can all be achieved in the same operation; to apply the work learned is a good consolidation and may also be an evaluation of whether it had been successfully learned in the first place.

Application   'Why are we learning this?' is a very common question, one to which the new teacher would often like to reply 'I haven't a clue, except that it is in the syllabus'. From the point of view of motivation, this leaves the teacher high and dry, resting his authority for tedium on a far distant examination, which may be inapplicable to many of the pupils. On the other hand, if knowledge can be applied and some reason given for its presence, life is more pleasant all round, and here the biologist is teaching the best of all subjects; there is little which cannot be made significant and relevant to the pupils. In a traditional syllabus, the relevance will have to be sought by the teacher, but pollination, germination and soil are all of relevance to the price of food in the shops.

Consolidation   A surprising number of new teachers think that once through the cerebral cortex of the pupil is enough, and that the information delivered then goes straight into the long-term memory store. It is of little help towards consolidation to ask the class what it remembers of last week's work; repetition is not consolidation, it may well be the repetition of wrongly understood work. Application is better, whether it is the application of knowledge to a new set of questions or to a real situation. If the first impact of the new work has been, for example, by practical work, then it is preferable to consolidate using a different method, for example a film loop, or making a model.

Remember also that *verbal* consolidation by several methods, such as a group filling in its worksheets together, is a useful method provided the lazy ones are not just copying – then it isn't going into the cerebral cortex at all, they might just as well be taking dictated notes. Classes which are used to dictated notes always resist strongly any change in method – they have to *think* for a change! 'But they must have something in their books' – yes, but preferably in their own words, which must first be in their mouths and then in their heads, not going straight through their ears to their pens without touching the brain at all. One stage up in the hierarchy is gap-filling, where most of the verbal work is done by the teacher, or simple multiple choice questions. Then comes writing short sentences, answering simple questions, then writing a series of answers to questions which give a full account of the work. Last of all comes the essay, but beyond this comes the consolidation of the consolidation – writing memo cards for revision!

There are occasions when the whole class can make a common record, which is a good means of consolidation, and useful not only where English is not the first language of the class – and this is why it was first originated – but when the added motivation of seeing one's work in print is thought necessary. The work is divided into sections which are given to groups to debate the best form of answer and record of the work done. This is then presented to the whole class for criticism and the final version duplicated.

Evaluation
1 By non-verbal feedback from the faces of the class.
2 By verbal feedback as work and worksheets are discussed.
3 By completion of work, by tests, by application.

This is how the teacher evaluates the effect of his learning situations – but the pupils need to know how they have done also, and they need to feel the satisfaction of completing a finite job successfully. Work that has been well done deserves to have attention paid to the marking – but a teacher swamped with marking dictated notes does not have the time to mark evaluative work, so beware. The more responsibility the pupil takes for basic record-keeping and the more time the teacher spends on finding out whether the work has been understood, the more efficient the situation.

Further considerations for lesson planning

Motivation

Lack of motivation is one of the biggest contributory causes to lack of 'discipline' and one of the worst offenders here is the topic of soil, although osmosis and germination run it a close second.

In such topics try to answer these questions:

Lesson planning

(a) What is the reason given for including this topic? All too often the answer is 'because it's in the syllabus'.
(b) Can I make it significant to the class; is there any relevance to everyday life in it?
(c) If the answer to (a) and (b) is hopeless, then can a different approach be tried? Can I use a problem-solving approach? Can I use it from the point of view of criticizing experimental design? Or must it just be a test of persistence and determination, with some kind of reward for satisfactory completion?

Class activity

We all get bored if we do the same thing for too long. So look at your lesson plans in terms of pupil activity. Consider the age and ability of the class – and their likely attention span.

Am I going to talk for too long? (the commonest fault)
Are they writing too much?
Do I need to show all of the 35-minute film?
 and so on.

A change of activity is a critical point in a lesson. A smooth change from one pattern of activity to another is essential for a well managed class. So consider:

How will I switch into the practical work?
What will I say to get the class looking at their books/posters/slides/ demonstration, etc?

Language

Since the publication of the Bullock report, *A Language of Life*,[94] the staff in schools have become much more conscious of the importance of language – both written and verbal.[66, 81] Language is too important to leave to the English Department. When you are planning your lesson consider the part that language can play. Are you introducing any new terms? Can you break the word down into constituent parts to help both understanding and the memory? Words such as exothermal, photosynthesis and epidermis are good candidates here!

You could start with a word bank. Write the 'biological' words on cards, with the definition on the back. Then put the card up on the wall when you first use the word. Keep the cards in a box file, situated so that the pupils can easily use them for later reference.

If your class is going to use printed materials – books, leaflets or even your own worksheets – is the language appropriate? Recently there has been an increased interest in the *readability* of textbooks.[10, 24, 35, 48, 50] There are various, quite simple ways that you can discover the approximate reading age of written material. You might like to test a standard textbook yourself. Don't think, though, that readability is the only factor determining whether a child can read and understand. Layout, design, illustration, the ways in which new words are introduced,[22] and perhaps above all motivation, are all important factors.

Consider what writing you are going to demand of the class. You shouldn't rely only on the formal presentation of the language of science. [22, 63, 102, 115] For one thing it's hardly the most exciting language to widen the horizons and stimulate the interest of your classes. Find ways of giving your classes opportunities of using other forms of language to report or comment on their biology. Here are some suggestions:

a poem about the behaviour of a woodlouse
a letter to an uncle explaining why he shouldn't put so much insecticide in his garden
a diary, the day in the life of an owl
write a composition on 'ride a red blood cell around the body'
a letter to a friend telling him or her how your reaction time was measured

'With all the insecticide that you use, Uncle George, by 1990 your prize marrows could become biological time bombs.'

Lesson planning

Concepts and behavioural objectives

If you look back to table 1 (page 24), you will see an asterisk marking 'variation' – this is a concept: crosses mark behavioural objectives such as observation and design of investigations. Particular bits of factual information alone rarely make a good lesson analysis; ideas can be consolidated by the application of concepts and by the use of behavioural objectives.

'Variation' is a good example of a concept too often taught by definition only and solely applied in academic situations rather than to practical realities. The medical and paramedical professions get very cross when patients deviate from the norm; they are not supposed to do so! And one wonders, for example, how much work on the evolution of man falls foul of non-application of variation – was Nutcracker Man a genuinely different entity or was he just a hominid at one end of the continuum of variation between very large teeth and very small ones?

Concepts are treated in detail later (pages 94–5); it is sufficient for now to be aware that if concepts underlie a piece of work, it is often useful to teach them consciously, preferably by application.

Similarly with behavioural objectives; if the topic is factually dull, see if you can use it to further accurate observation, criticism of experimental design, induction and deduction, literacy, numeracy and also self-discipline!

Even in curriculum reform projects, analysis on the aspects of motivation, concepts and behavioural objectives is important; one cannot just teach any text from page 1 to infinity without thinking about it systematically.

Revision lessons

Check through the following to see that you appreciate these points.
(a) As in all learning situations, pupils need to be *actively* involved. Where revision lessons are customarily occupied by either *not* listening to the teacher or staring at notebooks, pupils will resent having to *do* something.
(b) The main objectives of revision are to:
  (i) diagnose gaps in understanding;
  (ii) ensure that the pupils have *total recall*, not only recognition, of familiar facts.

  To fulfil these objectives, pupils need to try to write/repeat their work and check back on what they could not remember. They need to practise drawings and remember labels and experimental techniques.
(c) All pupils need *training* in how to organize their own revision work. Some prefer repetition aloud, some prefer diagrams, others prefer writing brief notes. All need help in providing the most efficient self-checking method for themselves. Teachers who provide self-check lists for each section of work should find revision easier.

(d) A secondary objective is to improve the pupils' critical ability, synthesis and application of facts to a new situation.

A method to try
Ad hoc 'Do revision with 4B' – with no previous preparation. Ask each pupil to set an exam question they would like to answer (also a useful way of setting a school paper!). You can then get competitive for a change and divide them into teams. Give marks to the pupil answering the question, *but* let the pupil asking it correct the answer in the first instance, and give her marks for correction. Answers should be telegraphic if verbal. If they are to be written, shuffle the questions and give them out – also get the answer written down by the setter!

Preparation for revision lessons
1 Inspection of recent examination papers – most schools have a supply.
2 Inspection of examiners' reports on questions in previous years – these can often be obtained by writing to the examination board.
3 Active revision with specimens – set a series of clues, with small workcards for each station, the answers to which make up the answer to a question. Check answers to small questions then ask them to write out essay questions. This gives confidence and helps them to answer fully.
4 Visual – use a loop or a film strip, and get everyone involved by giving a written quiz and pre-test to fill in – this diagnoses factual gaps and motivates to filling them up.
5 Key cards – these are actually sold! But it is a good beginning to ask classes to make their own on a certain area of the syllabus. This could be for homework, followed by essay questions to which skeleton answers can be written.

Psychological aspects
1 Examiners want to pass pupils – ask the pupils to help by careful answering of the questions! If course work assessment is used, give the pupils the idea of showing how wrong it was – in an upward direction, of course.
2 Total recall is impossible under examination conditions in biology (as opposed to mathematics, etc. – who gets 100 per cent in biology examinations?) so tell pupils to spread their time evenly over the questions.
3 Midnight oil burning may raise self-esteem by its sensational announcement to the peer group, but it is a poor investment.
4 Active revision is essential – *and give up when no more is going in.* It is a waste of time when intake is low; go and do something else, relax your tension, build up a good guilt feeling and then do another hour.

Barriers to efficient lesson planning

Lesson planning neurosis

If you wait till tomorrow, there is no guarantee that you will have a better idea for a lesson plan – only a bigger sense of panic because time is slipping by; this in itself hinders clarity of thought. Write it down today, even if it is unpleasing to your systematic mind – show it to another teacher for comment.

Biology lessons require organization of materials – are they available? Does someone else want them? Will they grow in time? Can we afford them? Popular films may have waiting lists of three months. The egocentricity of the teacher is also dangerous – many have believed that they will be the only ones needing the overhead projector at 9 a.m. on Monday morning, and have arrived at 8.55 a.m. to find it had already been booked, a week ago, by another teacher. Practical work needs at least a week's forward planning, films need longer, and then may never arrive at the right time for the pace of the class – eventually one hopes that they will be superseded by further developments which will allow for more flexible usage such as video cassettes.

Lack of resources

Most of this book assumes that your school or college has a good supply of animals, a well equipped laboratory and money for film hire and for buying materials. However, there are still many schools which have very little in the way of resources and those are in a state of dilapidation and disrepair. Your first question should be 'From whom can I borrow?' Teachers' centres, colleges and departments of education, and, for human and environmental biology, your Local Health Education Officer or Environmental Health Officer can prove useful in seeking out resources. On the other hand, if you are beginning teaching in a developing country and your luggage has not yet caught up with you, or conversely if an 'Act of God' prevents you from using the resources you had planned, here are some simple ideas to use as a basis for lessons.

Observation of the familiar from a scientific viewpoint For example, do not equate biological field work with a havoc-wreaking, conservation-defying day in the country – the bare bones of colonization can be seen on almost any city street, as the worksheet in Figure 2 shows.

The human body One teacher, at a conference which introduced a new science teaching project, was dismayed to find that she might be asked to teach human anatomy; she had, she complained, no skeleton, no models of the human body. She was reminded that she had at least twenty skeletons in her classroom and that even medical students begin by examining what they can see on the surface of the body, such as:

Colonisation

Can you fill in on this table an example you have observed?

WHAT HAPPENS	BAKER STREET EXAMPLE	MY EXAMPLE
Rock surface is rough/ is made rough	Limestone etched* by acid in air. (*Dissolves some of it)	
Water is available. Alga grows, maybe lichen	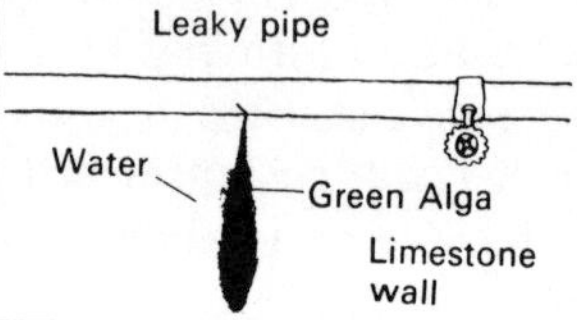 Leaky pipe — Water — Green Alga — Limestone wall	
Dust is caught. Moss grows. (How is soil formed? From what?)	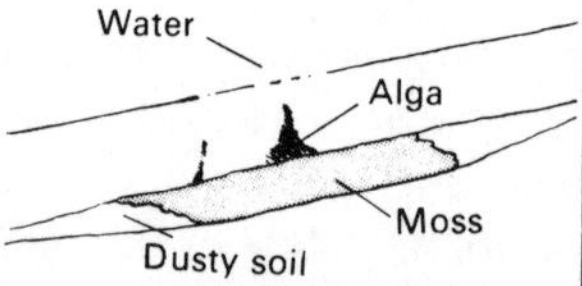 Water, Alga, Dusty soil, Moss	
Ferns — if plenty of water	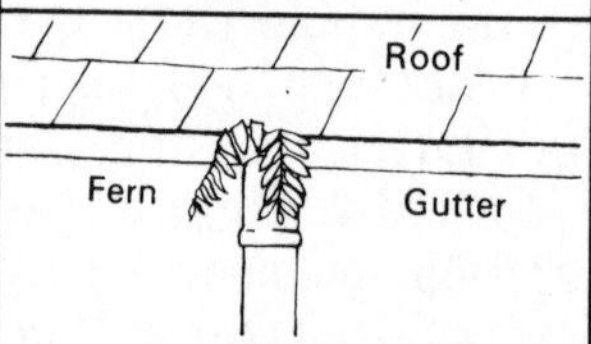 Roof, Fern, Gutter	
Flowering plants. (Why do plants need soil?)	 Sherlock Bellamy, Ragwort	
Trees	None there — do you know of a sycamore which is splitting a wall?	

Figure 2 *Colonization*

Lesson planning

'Quite remarkable, Taylor, but I think that will be enough for the moment.'

(a) Counting ribs, phalanges, etc.
(b) Determining limits of bones such as the femur, or the radius/ulna.
(c) Finding different types of joint movements – this often leads to displays by the 'double' jointed; that is, those whose ligamentous joint capsules are more elastic than the norm.
(d) Looking for veins – this leads to finding how much fat is covering them, and pupils can pinch up a skinfold, below the shoulder blade, to compare their subcutaneous fat layer with the adult norm of 2.5 cm. This is in fact a better determinant of obesity than measuring mass.
(e) Finding where an artery can be pressed against a bone – pressure points.
(f) Observing the *outside* of the eye – very few biology textbooks ever even mention it, being highly preoccupied with lenses and such like.

There are many more complex observations, which can be found, for example, in Nuffield Secondary Science, Theme 3,[78] such as the examination of the distribution of sweat glands.

Measurement Appendix B contains two extended worksheets on measurement of humans, and by no means exhausts the possibilities. For the teacher

Getting up like a biologist

Can you complete this table? There are no RIGHT answers — discuss them.

Stimulus	Sense organ used	Interpretation	Response
It's dark.	Eye.		
Bacon frying.	Nose.	What a lovely smell! I'm hungry!	Gets up.
Mum calls, "Time to get up".	Ear.	I can sneak another five minutes.	Rolls over.
Radio says there's a bus strike.	Ear.		
Remembers it's Saturday.			Rolls over.
		I don't feel like school today.	
		Blankets have fallen off.	

Figure 3

Lesson planning 37

It ain't what you say!

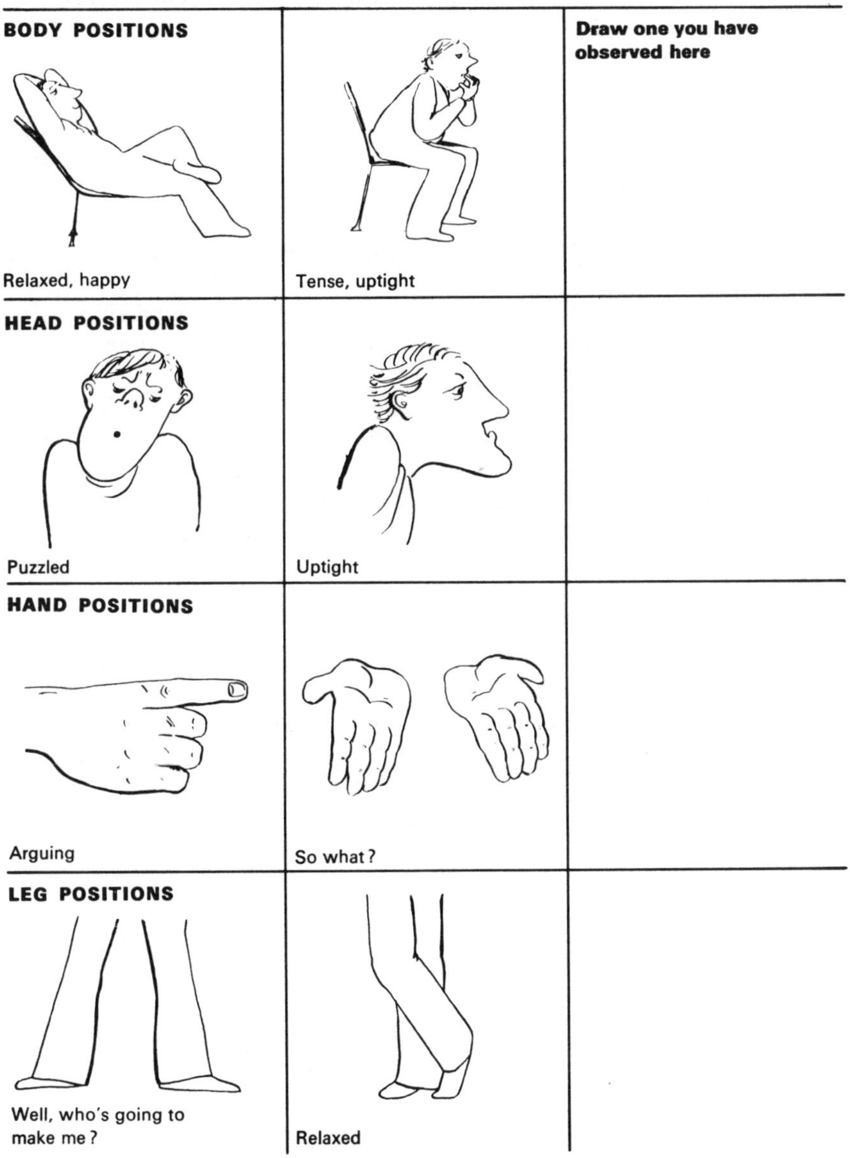

BODY POSITIONS		Draw one you have observed here
Relaxed, happy	Tense, uptight	
HEAD POSITIONS		
Puzzled	Uptight	
HAND POSITIONS		
Arguing	So what?	
LEG POSITIONS		
Well, who's going to make me?	Relaxed	

Figure 4

desperately short of resources, it is advisable to find the lengths of your own thumb joints, paces, etc. so that in the absence of rulers pupils can calibrate their own measuring devices and thus learn a healthy respect for their validity! Measuring rates at which hair and nails grow is a useful introduction to long-term experiments and their problems.

And with measurement comes correlation, as Appendix B shows.

Human behaviour Appendix B worksheet shows you how to begin with a simple pupil-made reaction timer – Nuffield Secondary Science, Theme 3[78] will show you how to develop the work and show its significance and relevance for the car driver, for example. The worksheet in Figure 3 demonstrates how pupils can begin to observe their own behaviour – back to scientific observation of the familiar.

And when in doubt, one can always use non-verbal communication studies (Figure 4), a wonderful way to acquire a quiet class!

Very simple apparatus, very simple models There was once a biology teacher who actually made the apparatus and did the dissection necessary for the demonstration of a Stannius ligature on a frog's heart in a Japanese prisoner of war camp. While such pinnacles of achievement are not possible every day of the week, the scavenging frame of mind is to be recommended to all, and the pages of science teaching journals and elsewhere[57] abound with excellent suggestions for apparatus made from junk. Don't neglect the UNESCO source book![109]

One of the simplest pieces of apparatus is the clinical thermometer, and Appendix C is devoted to information and suggestions on its use – examining the reason why one shakes the mercury down, measuring toe temperature, for

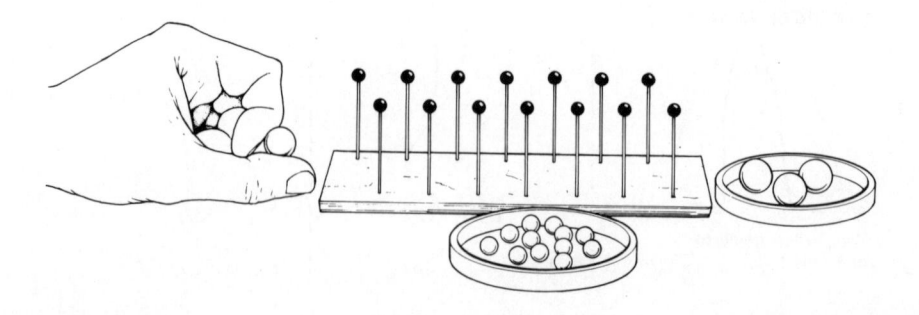

Figure 5

example, as well as that of mouth and armpit, measuring temperatures at different stages of the life cycle, at different times of day, at different external air temperatures, etc.

Figure 5 shows one of the very simplest models, used in an introductory session on the working of the kidney for fourth form boys of average ability. The pins were stuck in a dissecting board, and the balls were caught in Petri dishes. It gave rise to some good discussion on the accuracies of models, because the boys discovered that if the balls were flicked hard enough, even the smaller ones got across the board – 'Well, it's high blood pressure, Miss, isn't it?'

The major barrier to this kind of work is the attitude of mind which thinks it can't do science without a proper laboratory and proper apparatus. Often pupils will think more about the simple stuff, merely because they suspect there must be more to it than appears on the surface! Whereas well-stocked laboratories and elaborate apparatus may produce the contempt which familiarity is said to bring.

The kitchen While not expecting the sophisticated kitchen equipment named in Vicki Cobb's book *Science experiments you can eat*,[27] you just *might* find a friendly and co-operative home economics teacher in your school who could let you have some resource material. Don't forget, however, the general safety rule that nothing used in the laboratory should be eaten – and that applies to sweet-eating as well! If co-operation is not to be found, strategies for teaching can be seen in the most unlikely situations. For those starting teaching in tropical schools your pupils can tell you what is found in the local market, and you can, once again, help them in their scientific observation of the familiar.

Nuffield Chelsea Home Economics : A Scientific Approach is a current (1980) project working at the interface of these two areas of the curriculum. It is to be published by Hutchinson and the Nuffield Chelsea Trust.

3 Mixed ability teaching

'The work is combined in that study,' said the chaplain. 'Stalky does the mathematics, M'Turk the Latin, and Beetle attends to their English and French. . . .' 'It amounts to systematic cribbing,' said Prout. . . . 'No such thing,' little Hartopp returned. 'You can't teach a cow the violin.'

R. KIPLING
Stalky and Co (1899)

If you intend to teach only the top 20 per cent of the school population, you may skip this bit; most of them are bright enough to learn anything in lower school biology, no matter how badly you teach. You can comfort yourself that your bad teaching will at least make them learn how to learn by themselves, which will stand them in good stead when they reach university.

But even when the top pupils of the top stream reach university and graduate from it in biological sciences, can you honestly say that such a streamed group, such a selected élite is of equal ability? I suppose it is statistically possible but highly unlikely to occur. So perhaps you had better examine the reasons why, for example, biology graduates on a teacher education course need to be taught as a mixed-ability group; it might help you to realize that the lack of understanding in the examination classes you teach could be due to your lack of awareness of their mixed abilities rather than to their innate ineptitude.

'We once divided our first year into two and kept one half mixed while streaming the other. The experiment yielded the information that the children with the best teachers made the most progress.' (Mixed ability teaching. A report on a survey conducted by the Assistant Masters Association. 1974.)[6]

It is possible that the best teachers have always used mixed ability methods; most experienced teachers known to me will learn absolutely nothing from this section, although they may never have attached a label saying 'Mixed Ability Expert' to their teaching.

What do you mean by mixed abilities?

Intellectual

Learning pace

Individuals learn at different speeds.
Individuals learn different subjects at different speeds.
Individuals learn different parts of one subject at different speeds.
Individuals' speeds vary according to time, place, physical health, home worries, teacher, the outcome of the local football derby, what they had for breakfast, etc. ad infinitum.

Nevertheless people still like to kid themselves that they can teach 25-plus individuals at the same pace!

While classes continue to be grouped by chronological age (relevant to bureaucracy rather than to education, but then someone has to organize the business) they will (if one but applies the basic concept of variation) form a continuum of ages of mental development, and so be of mixed ability.

One answer to this dilemma lies in individual learning situations, where individuals can work at their own pace, and get individual help from the teacher at the time and at that point in their learning when they need it.

Methods used

Worksheets, programmed learning, team teaching of basic material followed by small group work for consolidation. Plus the realization that you cannot ride a horse at a gallop all the time – sometimes horses want to go slowly, sometimes just roll around and kick their legs. It takes experience to ensure everyone is working to capacity in a large class.

Warning: Individual learning situations are the basis of mixed ability teaching, but like any other method, if used all day, every day, boredom sets in. Chapter 4 of this book deals with the variety of methods which need to be used.

Those who finish first

One of the ungrounded fears of mixed ability teaching is that it 'keeps back' fast learners. In fact these are more likely to waste time and become bored in whole-class teaching, geared, as it must be, to one learning pace.

Teachers should ask these questions when a pupil finishes long before the rest.
1 Is this work well done or rushed?
2 Would it benefit the fast pupil to help tutor the slower pupils?
3 Would it benefit the fast pupil to be given a piece of work of high status/ high entertainment value? For example, materials which are too expensive for a whole class could be used by the fast finishers and demonstrated to the

Table 3 *Overview   Which abilities are mixed?*

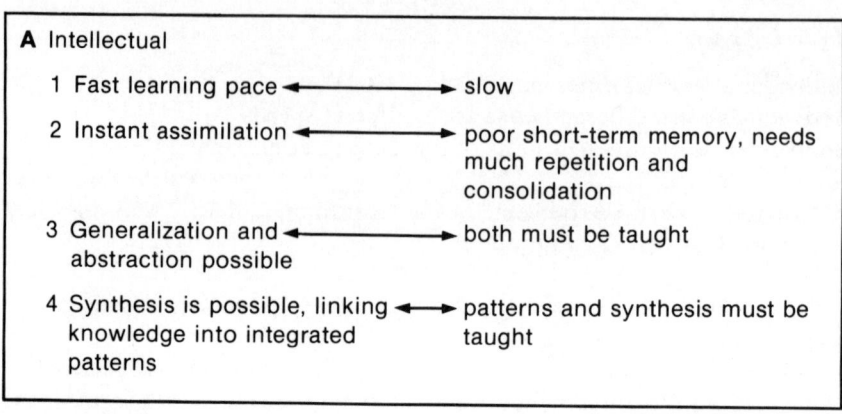

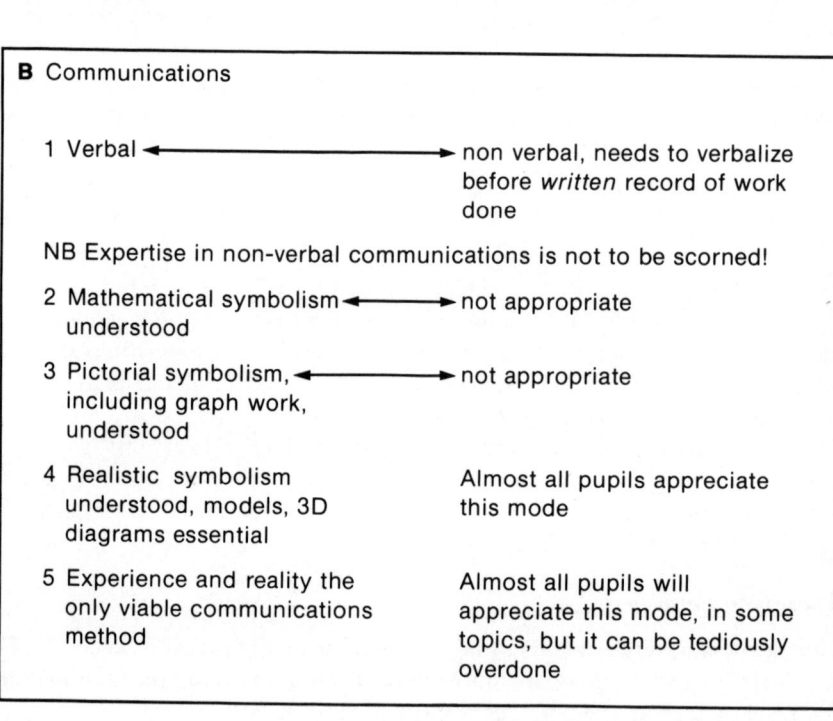

A Intellectual

1 Fast learning pace ←——→ slow

2 Instant assimilation ←——→ poor short-term memory, needs much repetition and consolidation

3 Generalization and abstraction possible ←——→ both must be taught

4 Synthesis is possible, linking knowledge into integrated patterns ←——→ patterns and synthesis must be taught

---

B Communications

1 Verbal ←——→ non verbal, needs to verbalize before *written* record of work done

NB Expertise in non-verbal communications is not to be scorned!

2 Mathematical symbolism understood ←——→ not appropriate

3 Pictorial symbolism, including graph work, understood ←——→ not appropriate

4 Realistic symbolism understood, models, 3D diagrams essential — Almost all pupils appreciate this mode

5 Experience and reality the only viable communications method — Almost all pupils will appreciate this mode, in some topics, but it can be tediously overdone

C Skills

1. Competent reader ↔ slow
2. Comprehension (given that communications methods are appropriate) ↔ little understanding
3. Logical thought (expressed in terms of action as well as words) ↔ lack of critical appraisal of evidence and inference
4. Competence ↔ technical ineptitude
5. Draws well ↔ incompetent

D Social and Cultural

Is motivation likely to be by

(a) Short-term aims, e.g. feeling of success in self, leisure and jobs
(b) Long-term aims, e.g. exams, professional aspirations
(c) Technology (most boys)
(d) People, especially children (most girls)
(e) Authority, and repressive discipline
(f) Autonomy and self-discipline

E Assessment

(a) Self checking by pupil
(b) Assessment of effort as well as attainment
(c) Assessment of skills and application of knowledge, rather than merely recall

BC Communications

from pupil

(a) Non-verbal
(b) Verbal
(c) Mathematical
(e) Pictorial
(f) Technical
(g) Demonstration, concrete, reality

rest of the class. Three-dimensional models which can be made from kits are another possibility, or the pupil can design his own.
4 The pupil really *might* like to begin his homework or continue his project.
5 The pupil might be allowed free choice of any apparatus in the laboratory. This makes a good end-of-term lesson for all.

Graded worksheets are another method of ensuring that the basic work is done by all, while the faster paced can go further than this in application and some problem solving. Such graded work is also used in some forms of assessment, which is its right and proper place; too many graded worksheets could make the slower pupils doomed forever to repetition with little application.

Short-term memory

Teachers who *admit* to having a poor short-term memory take careful notes of times, dates and arrangements as well as full lecture notes; they generally cope with life better and achieve more than those who are ashamed of their handicap, and pretend it does not exist!

Short-term memory for what? To some, mathematical symbols and numbers are easily assimilated, to others visual stimuli in colour are taken in most quickly, some remember whole pages of textbooks, others are best when observing human actions. The average secondary school pupil, however, cannot watch a long demonstration and remember every move made; this seems to be still a favourite teaching method in some subjects, but is now not generally favoured in biology. Biologists, however, do tend to show long films (some of which are superb) without means of consolidation and repetition; little will even reach a short-term memory already saturated with TV watching and even less will end up in the long-term memory store.

With least able pupils, facts and simple techniques which seem to have been learned one week will have disappeared by the next week; they may have been assimilated by the short-term memory but have not been passed on to the memory store. Teachers of remedial pupils need saint-like patience to repeat and repeat work, and also need professional training in how to do it. Verbal repetition itself is of little use, and consolidation must be by other methods – visual, practical application or through experience, for example.

It has been shown that it is unwise to lecture to graduates, without a change of demand on attention and short-term memory, for more than twenty minutes; with the least able pupil you are likely to teach, one minute is a more realistic time, after which the memory must be supported by a variety of methods, as outlined in the next section. 'Just to "do science" for a double period each week may be merely to provide them with long enough intervals in which to forget.' (Newsom report, paragraph 429)[93]

Generalization and abstraction

Prep-schools* have known for a long time that mental maturation sufficient for pupils to deal with abstracts and generalizations rarely begins before 13 years. Piaget[11] had some words to say about it too and described it as the ability to deal with formal mental operations rather than concrete (real life, or at least realistic models of it) experience. The least able pupils you will teach may *never* be able to form abstractions and generalizations for themselves; if you therefore do not provide them, or the means for their formation, much of biology will mean little. Circuses, more fully described in Chapter 4, are means whereby pupils can *experience* concepts and generalizations and discriminate between, for example, instances and non-instances of osmosis or energy transfer. There is little point in teaching a verbal definition of osmosis if it cannot be applied; not that this will matter very much in an examination – but it could matter to a pupil who will be a pathology laboratory technician later in life. Many average pupils will be able to apply concepts they have learned by experience without ever being able to verbalize them; many examination successes will be able to verbalize but not apply. It seems incongruous that ability in application *without* verbalization is not rewarded by an official piece of paper.

To quote Schools Council Working Paper No. 1 (1965),[95] now long out of print, which dealt with science for the young school leaver:

'This would involve taking a good deal less of the learning process for granted than is often the case at present. Relationships upon which significance will depend, which may arise spontaneously in the minds of their abler fellows, will need to be made abundantly manifest to Newsom† pupils, particularly to those of lower ability.

Attention certainly needs to be given to ways of developing basic concepts of fuller significance. This is a process which is all too easily hurried, with the Newsom pupil who commonly finds it easier to think in concrete rather than in abstract terms and who comes somewhat slowly to a generalization or theoretical idea. Without the necessary repetition and variety of experience a term can easily lack significance, and if the pupil has to use it without this adequate understanding his means of communication begins to break down and he experiences failure and frustration.'

The somewhat condescending term 'Newsom pupils' must be qualified on the ground that 'average' ability was and still is measured with reference to academic attributes only. There are many examination successes who have suffered because the learning process has been taken too much for granted,

---

* Preparatory schools in the UK which prepare children for entrance to fee-paying schools at 13-plus.

† This was derived from the report *Half our Future*[93] on the education of pupils in Britain aged 13–16 of average and less than average ability, which was chaired by the late Sir John Newsom.

and not a few biology graduates who would have benefited from circus teaching such as that on energy transfers which was originally designed for 'the lower streams'.

Synthesis

To quote Working Paper No. 1 again:

'Lastly, it is important for the Newsom pupil that his science as a whole should have some significance of pattern and ultimately "add up". "The field of science is so wide that what is done in schools can jump from one facet of the subject to another without much sense of cohesion developing" (Newsom, paragraph 426) and this is a danger that must be consciously avoided. This will happen, however, only if the general pattern is clearly envisaged by the teacher from the outset.'

Unfortunately the teacher very often has little idea of patterns either! Most theories of the learning process indicate that before ideas are stored in the long-term memory they need to undergo organization of some kind; various authorities talk of pattern, schemes, field cognition or frames of reference, but, whatever learning theories are favoured, if teachers do not give some simple basis for organization of facts and principles then it is manifestly ridiculous to expect any but the most intellectual pupil to do it for himself.

Help is at hand, however, but only for the 13-plus pupil. SCISP, the Schools Council's Integrated Science Project[97] based the organization of its work on Gagné's principles[45] and on 'patterns'. It not only provides the basic patterns but also links the patterns together. Incidentally it also implements another recommendation put forward by the Newsom report, that of linking allied subjects with biology, chemistry and physics. The Newsom report specified mathematics and geography, SCISP also provides social sciences; it stops short of poetry, although the Newsom report, paragraph 438, states: 'It is not unthinkable that the notice board in the science laboratory should sometimes display a poem; poets as well as scientists are observers.' The Nuffield Revision (1974) O-level chemistry leaflet[80] on 'The Halogens' *does* refer pupils to a poem! ('Dulce et decorum est' by Wilfred Owen.)[46]

And to quote Working Paper No. 1 again: '... it is clear that the problem of dealing with personal, social and moral implications when they arise in connection with science will have to be faced. No treatment of science with these boys and girls can, in fact, claim to be realistic or of adult stature if it excludes or sheers away from issues of this kind and much that may be of vital interest to the pupils is likely to be in this category – very much at the interface of the traditional school subjects, all with their own vested interests.'

After all this the reader might well think that SCISP, providing patterned integrated science, is for 'Newsom' pupils – not so; those who follow SCISP take an examination which will award them the equivalent of two O-levels.

Nuffield Secondary Science[78] was, however, originally for average and below average pupils; it is organized into eight themes, each of which occupies a large teacher's guide, not a textbook for pupils. It was extended to cater for pupils capable of taking public (CSE) examinations and it has been used with academic (O-level) classes who also appreciate organized knowledge and a variety of method.

Finally on the subject of synthesis, recording work done and knowledge gained is an essential part of pattern-forming as well as of the consolidation of separate facts.

Communication

To many biologists, mathematics is an inappropriate symbolic language, and the passage on Giky Martables (page 7) shows that the terminological exactitude of biology can also be totally lacking in communication for many pupils. It is possible that some non-verbal pupils escape with delight into mathematics; this is not to say that they are less able, some of my best friends are mathematicians and really quite bright! So in any group of people, whatever their intellectual capacity there will be mixed abilities in communications; hence any professional teacher needs to master a variety of methods of communication.

Verbal ability

Learning from listening to what a teacher says is not so effective as learning by seeing things done, which in turn is less effective than learning by doing things. Where verbal ability is high, teacher talk will be effective, but where it is low, attention wanders, frustration at not understanding occurs, and other methods are preferable.

In particular, pupils with low verbal ability need to verbalize new learning before they can think about it or write about it. Hence the emphasis on *small group work,* possibly using worksheets, where pupils educate pupils by putting thoughts into words and actions. If the teacher talks most of the time or worksheets are filled in in silence, learning is less effective and education in communication, which has a great validity outside the school as well as in, is neglected. Some might argue that if you have good examination qualifications it does not matter that you are inarticulate – there seems to be little evidence for this point of view, and verbal education is as essential for the scientist in the university as for lesser mortals.

Figure 6 is a worksheet from *Quest,** showing conscious (and effective) education of verbal ability.

---

* *Quest* is an integrated science course. The initial work was produced by Peter Herbert, Elliot School, Putney; John Merrigan, Christopher Wren School, Hammersmith; and John Lewis, ILEA Science Advisory Teacher.

002A   QUEST   00201 A/1

Sticky liquids ....

Look carefully at bottles A to H.

Write one sentence to describe each of them.

Answers expected:

A	Is a thick oily liquid	(liquid paraffin)
B	Is a bottle of yellow sand	
C	Is a clear watery liquid	(e.g. water!)
D	Is like pink coloured water	(e.g. cobalt chloride solution)
E	Looks like crushed ice	(2-methylpropan-2-ol - freezes at $15°C$)
F	Is a dark brown, oily liquid	(crude oil)
G	Is a browny orangy solid made of beads	(ion exchange resin)
H	Is a number of silvery balls	(3 mm ball bearings)

Here are eight descriptions. Each fits one of the liquids.

         syrupy

         runny black

         runny silvery

         pale brown

         watery

         a pink colour

         like sand but more shiny

         like wet ice

Fill in each space with a description to fit each liquid. Use each only once.

Answers expected:

A syrupy, B pale brown, C watery, D a pink colour, E like wet ice, F runny black, G like sand but more shiny and H runny silvery.

© 1974 QUEST

Figure 6

/ Mixed ability teaching

Working in small groups is, like individual learning, a basic priority in mixed ability teaching method whether in university or in comprehensive school. Dividing classes into fast and slow groups is a kind of streaming in miniature, emphasizing only intellectual skills; but where the hamfisted intellectual is helped by the non-verbal pupil with lots of technical skill and knowhow, where the girl who writes well is helped in her mathematics by the boy whose English homework she does in return, and where the artist who can't count helps the computer science whizz-kid who can't draw, all gain, not only in the amount they learn, but in respect for each other's attributes and acceptance of their own limitations. Co-operative effort in the classroom and the education of pupils by pupils also often produce higher standards of learning and less tired teachers, besides what it does for socialization.

Mathematical symbolism

Give or take the odd bad teacher, I would suspect that the ability to communicate using mathematical symbols is genetically determined; if you are born without the appropriate circuits in your brain, mathematics will continue to be little but a feeling of failure, however hard you try, when only symbols are used. There is an alternative hypothesis which says that the preference for such symbolic communication is a neat way of escaping from parents who talk too much!

We have already examined the necessity of helping pupils to learn the terminology of biology. Pupils also need to be helped in the use of mathematical symbols, and merely to say 'Let sigma be the sum of all the numbers' means that the teacher is taking too much of the learning processes of the general population for granted. Nowhere more than in mathematics does knowledge need to be organized into hierarchies, nowhere does the learning process have to be so carefully organized.

Biology-mathematics is much more difficult than mathematics-mathematics – unless it is visual, such as a graph. It also seems pointless to teach statistics in genetic studies unless calculators are available, if not more sophisticated electronic aids. The usual approach via a load of mechanical arithmetic is designed to put even the most able off standard deviation. The recognition of this fact in schools using mixed ability teaching (for example, leading in from a study of the different prices and weights per packet of different washing powders) has brought in the use of computers and calculators to the advantage of all.

The worksheet in Figure 7 is one teacher's attempt to reduce the terror of mathematical symbolism; the arrows were added when another teacher was explaining the worksheet to another class – do they help or hinder? However imperfect this worksheet, it is a considerable advance on the presentation of a formula into which (as into a sausage machine) various numbers are inserted and a result produced, which may be correct, but which may represent little or no understanding of the matters explored.

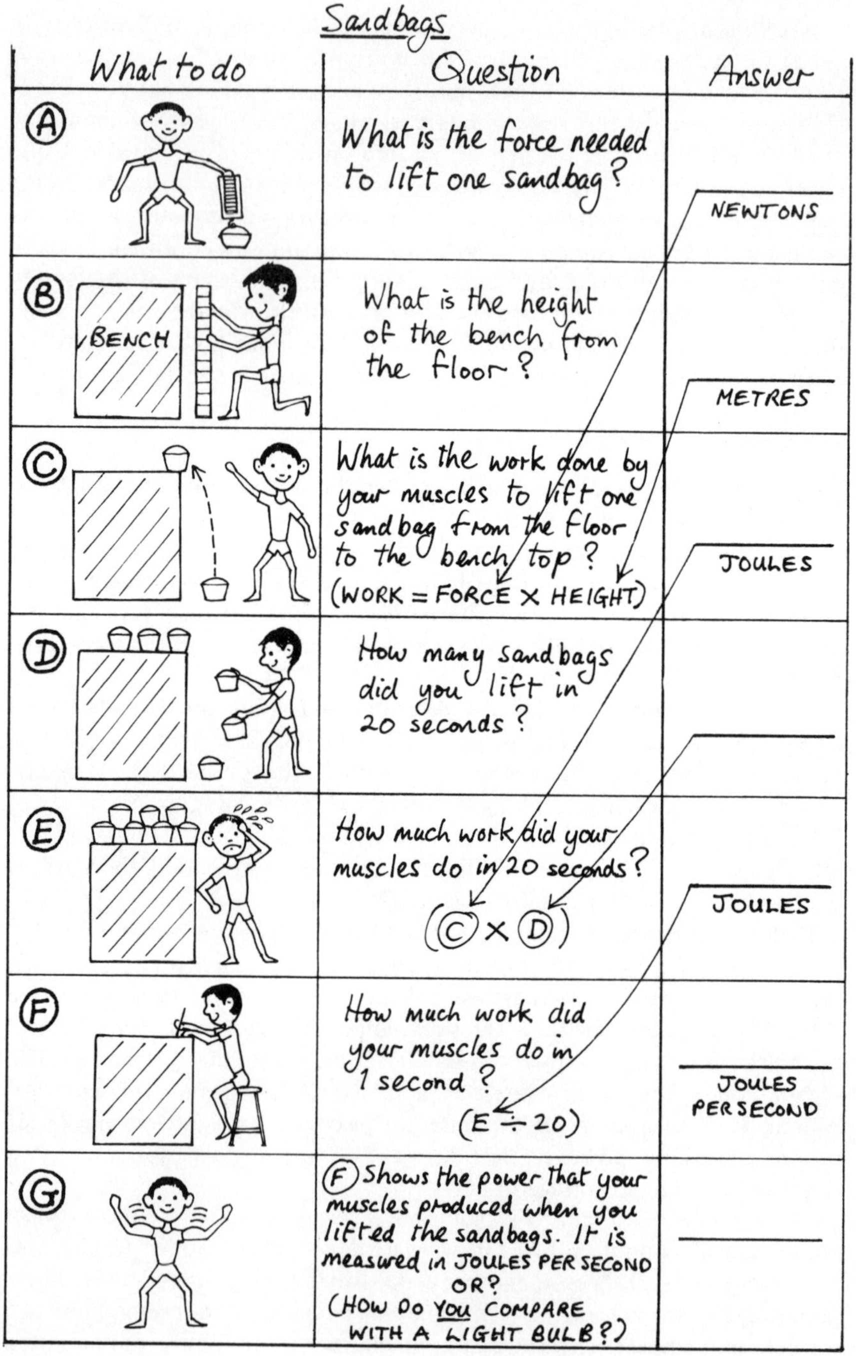

Sandbags

What to do	Question	Answer
Ⓐ	What is the force needed to lift one sandbag?	NEWTONS
Ⓑ	What is the height of the bench from the floor?	METRES
Ⓒ	What is the work done by your muscles to lift one sandbag from the floor to the bench top? (WORK = FORCE × HEIGHT)	JOULES
Ⓓ	How many sandbags did you lift in 20 seconds?	
Ⓔ	How much work did your muscles do in 20 seconds? (Ⓒ × Ⓓ)	JOULES
Ⓕ	How much work did your muscles do in 1 second? (E ÷ 20)	JOULES PER SECOND
Ⓖ	Ⓕ shows the power that your muscles produced when you lifted the sandbags. It is measured in JOULES PER SECOND OR? (HOW DO YOU COMPARE WITH A LIGHT BULB?)	

Figure 7

Pictorial symbolism, including graphs

Graphical methods, for example of converting Fahrenheit to Celsius, often seem to be more easily used by more pupils than numerical methods, as these examples from Revised Nuffield Biology (Text 1)[79] demonstrate.

What is needed is a simple way of converting measurements made in one system of units (inches, for instance) into equivalent measurements in another system of units (metres). There are a number of ways in which this can be done. Two are given in figures [8] and [9], each concerned with converting degrees Fahrenheit into degrees Celsius, but these methods can be used to convert any equivalent measurements.

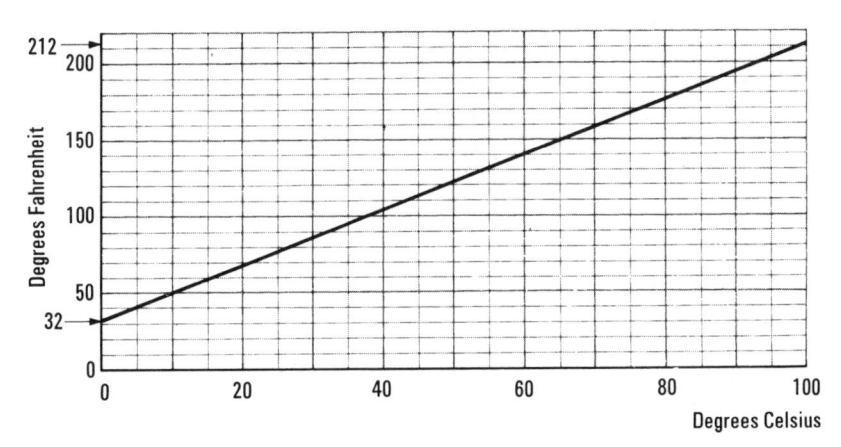

Figure 8

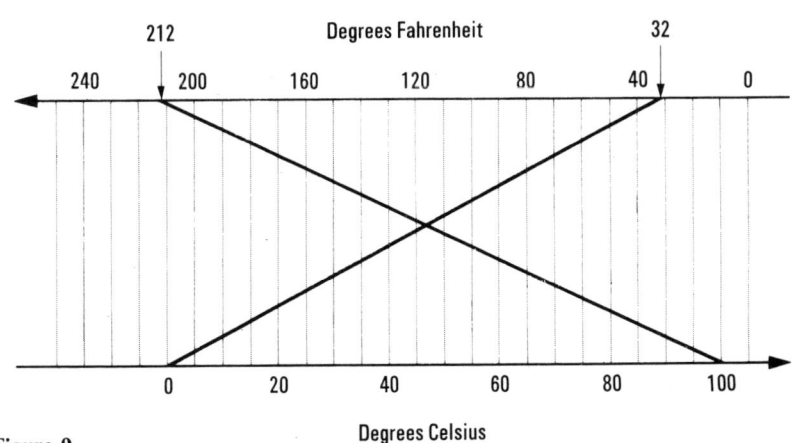

Figure 9

Use figure [8] to answer the following questions.
(a) The body temperature of humans is 98.4°F. What is the corresponding temperature in °C?
(b) Comfortable room temperature is 20°C. What is this temperature on the Fahrenheit scale?

Use figure [9], a nomogram, to answer the same two questions asked above. Compare the results obtained from the graph and from the nomogram. Which method do you prefer? The most accurate result is obtained when the graph and the nomogram are drawn large and with care, using a sharp pencil so that all the lines are fine. They must also be straight.

It can be tentatively hypothesized that pictorial symbolism is even more general as an understood method of communication.

Take, for example, the simple basic idea that the molecules in a solid have a more restricted movement and can pack into a smaller space than molecules of a liquid or a gas; it is essential that this be understood before work on density, diffusion and osmosis can have much meaning. To the pictorially literate pupil a simple two-dimensional diagram may suffice.

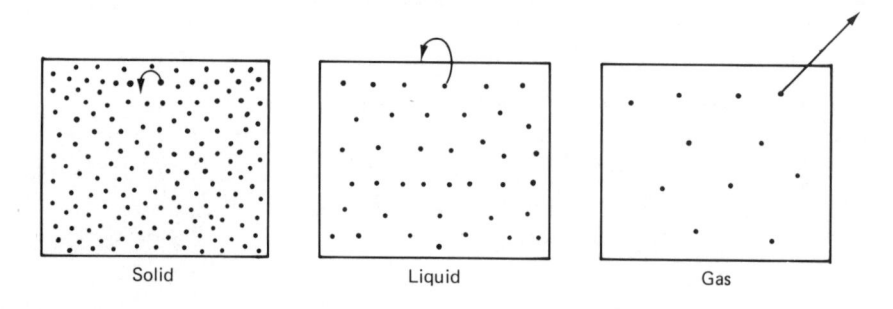

Solid   Liquid   Gas

Figure 10

Realistic symbolism

There are many to whom two-dimensional symbolism is difficult, and for whom a three-dimensional approach is needed. For these diagrams must be replaced, where possible, with a model. For the problem given above, perhaps the simplest model is to enclose the pupils within a chalk circle; ask them to run on the spot, representing the solid, then each to run in a small circle representing the liquid, counting how many of them overflow the chalk circle, then to run freely within the circle, whereupon the 'gaseous molecules' overflow all over the place. Any biologist who has attempted to interpret electron microscope sections will know how even the highest intellect appreciates a three-dimensional approach.

Biological models however have their dangers; some quite expensive ones show clearly that their makers themselves have difficulty in constructing a

Mixed ability teaching

'And now, Richards, I'm going to give you a simple demonstration of 'Human Power Output' - incidentally, best measured in watts.'

three-dimensional perception of a two-dimensional textbook drawing! All models and analogues whether pictorial, verbal, experimental or mathematical should have a critical appraisal of how they are *not* like the real thing appended.

Experience and reality

When following the worksheet on page 50 on measuring human power output, the pupils were experiencing forces, so it was just as well they were working in SI units rather than in the dear old erg which can only be experienced by an ant scratching itself with a hind leg (or so my physicist friends would have me believe), whereas the newton comes within the range of the human senses. For all pupils in some subject or some aspect of the work, experience and reality will be the only appropriate communicating method. It need not only be the application of knowledge which requires the use of a real situation; experience and reality can be the best way of introducing entirely new work.

Real situations can however be tedious, not only for the highest ability levels but for all, and the horrid emphasis on everyone having to discover everything in a real situation is luckily becoming unfashionable. Gerald Thompson's film on the Cabbage White butterfly[132] (Oxford Scientific Films) was criticized by a biology teacher because it took away the joy of pupils discovering the real thing for themselves. Since the discovery of the Cabbage White butterfly in the real situation requires hours of careful observation and the use of some very expensive lenses, most classes 'discover' a few tattered elderly specimens, some pickled caterpillars and maybe some real eggs on a real cabbage leaf. Thompson's expert perception of, for example, the eggs adds new dimensions through the *extension* of the pupils' sensory perception, and this is made possible by the film situation for most pupils.

Skills

'What are the essentials of the scientific method? To see, to wonder why, to attempt explanations, to test these by taking a closer look, is a common enough sequence of experience. The scientist repeats this process deliberately and in a controlled situation, learning to look closely, record accurately and say clearly what inferences have been made.' (Newsom report, paragraph 422)

It is not the purpose of this section to go into detail about the behavioural skills of scientific method save that it must be added to the simple outline given that *critical ability* needs to be encouraged through biology teaching rather than that the once prevailing mode of acceptance of dogma be perpetuated.

Reading ability

So, the reader reasonably asks, how am I to use workcards and worksheets when the pupils can't read? It is not expected that all new teachers would emulate the practice of the head of science (see page 8) who tape-recorded worksheets for slow readers.

There will always be pupils with reading difficulties which need specialist treatment, but today's TV-watching pupils, even those who will eventually enter university, need motivation to read. Here is a library lesson which attempted to motivate reading
(a) by introducing the books
(b) by asking direct questions which the books could answer.

This is how the teacher planned the organization necessary for a library lesson with a mixed ability class of 11-year-olds. How would your own circumstances differ? How would he mark the work, knowing that some are better at verbalizing than at writing?

Instruction to the class:
1 Arrange yourselves in pairs. [Teacher ensures a good reader is paired with a poor one.]
2 You are going to study the reproductive behaviour of any animal you are interested in.

Mixed ability teaching 55

3 To do this you will have to select a book which contains all the information you seek about your animals.
4 Choose your animal *now* – discuss it with your partner, quickly.
5 The books you may use are: Life books (Time/Life);
   books from the biology section;
   encyclopedias.
6 Only one person from each pair goes to select the book. The other person is to stay here and read the list of questions which will help you to find out and write about your particular animal.
7 You have only ten minutes in which to find your book.
8 Five will go to the Life books, seven to the biology section and three to the encyclopedias.
9 Your homework is to write an account of the reproductive behaviour of the animal you have chosen.

The teacher will:
(a) send pupils off to get books – decide arguments as to who goes where;
(b) give out sheets to pupils who remain seated;
(c) allow ten minutes, then they must all be sitting down and working;
(d) ask four or five of the pairs to read their work at the beginning of the next lesson.

This was a successful lesson – from the above list you can tell what kinds of motivation were being used. What advantages had this lesson over either teacher talking or use of film?

Comprehension
If the communications from the teacher to the pupil are appropriate, there should be comprehension; but to assess whether there is, effective communications must exist between the teacher and the pupil in one or more of the modes listed in the overview of 'Which abilities are mixed' (page 42).
(a) *Non-verbal* communication from the pupils is often the first indication that comprehension does not exist. This feedback is of prime importance in, for example, the lecture.
(b) *Verbal* ability to communicate exists in pupils, but to a variable extent; demagogues and question hoggers must be taught unparliamentary manners to allow the more serene members of the group to have their say, and teachers need consciously to strive to take verbal feedback from the whole group or class.
(c) *Written* evidence of comprehension needs to be structured in a hierarchy of increasing complexity, from simple gap-filling questions to those requiring answers in sentences and finally to the short essay. It may only indicate parrot learning, or fact copying from books, and needs careful organization if it is indeed to measure comprehension.
(d) *Mathematical* comprehension and communication by the pupil to the teacher – well, it does happen! For some pupils this is their preferred means of communication, but not for many.

(e) *Pictorial* communication is dealt with in Chapter 4 under 'Drawing' 78). It needs conscious training of the pupil in perception and symbolism.
(f) *Technical* communication by pupils should not be neglected as a means of demonstrating comprehension. Care needs to be taken that the pupil is not merely following a recipe but does understand what he is doing and why he is doing it. Hence it can rarely occur as the sole method of communication, although there are always some pupils who will prefer to show a teacher what happens if he does this to that, at the level of concrete operations, than to tell him about it.

Logical thought

The ability to succeed in the field of logical thought is not confined to the top 20 per cent of the ability range in the school population. Although if reports are to be believed, the top 20 per cent seems to suffer rather more than the rest from a plethora of the kinds of teaching which do not demand logical thought; problem-solving approaches, for example, have been condemned as taking too much time for exam-bound pupils. Since the exam-bound pupils produce most teachers, this type of practice will be perpetuated, unless determined efforts are made to break the vicious circle. Worse still, the other 80 per cent are so often subjected to watered-down versions of what is deemed right and proper for the top 20 per cent that one often wonders if logical thought in science today is the result of anything but genetics. Newsom, of course, said it all in 1963: 'Too much of the tradition of science teaching is of the nature of confirming foregone conclusions. It is a kind of anti-science, damaging to the lively mind, maybe, but deadly to the not so clever.' (Newsom report, paragraph 423)

Two pragmatic objections to the encouragement of critical appraisal of evidence and inference are:

Pupils who have been passively sitting in a laboratory for a few years, taking notes and jumping through the prescribed hoops provided by practicals and tests, resist a change to active involvement in lessons which ask them to take some conscious thought. Goodness! It cuts into their day-dreaming, fiddling and doodling time! Teachers therefore often report thankfully that the class does not like the new régime and sink back into the old.

Having sparked off a class to think and criticize, teachers may find at first that the barrage of questions is overwhelming – it is certainly the most tiring part of teaching, the constantly solving and communicating processes which accompany such a released flood. Teachers learn to direct the seekers to find their own knowledge, and find that the flood eventually abates. But why bother about logical thought as long as the pupils know their facts?

To whose advantage is it if schools produce pupils who have never exercised their logical thought processes?

Technical competence

Technical ineptitude may well be due to adolescent growth problems; bones complete their growth before the muscles necessary to move them grow fully, and lack of co-ordination, particularly of fine movement, may, therefore, be only a passing phase. With some of us, however, it lasts for a lifetime! Success in mastery of simple techniques in biology is pleasant, but teachers need to be aware that in spite of the greatest possible effort, some pupils will rarely achieve much in this field.

Conversely there are many pupils who can achieve a good deal in this field if careful thought is taken in the sequential planning of skills in learning, just as in factual learning. This applies no less to skills than it did to generalization and the ability to organize knowledge. I remember being appalled, as a precocious 11-year-old, when we spent a whole English lesson on the use of the dictionary – I simply could not understand why anyone needed to be taught such a thing, having used them happily for years without any teaching of the method needed. Like all fast-paced learners, I was similarly appalled when I had to *teach* pupils how to read scales on spring balances, on ammeters – and when it came to the clinical thermometer, I was forced to take very much to heart Working Paper No. 1's words about taking less of the learning process for granted! The steps involved here are:

Use of appropriate vocabulary about temperature, warmth, etc. There is little point in giving instructions about skills in an elaborated code, before giving the pupils the opportunity to acquire it; refinement in the choice of adjectives is often an important part of learning skills.

Ability to read scales; the variety of scales used on thermometers is considerable, and one finds that pupils, when transferring from one scale to another, often do not realize the difference.

Ability actually to *see* the mercury in a clinical thermometer – some provide a pair of guidelines, showing where to look and hence at which angle to hold the thermometer to the eye.

I am grateful to the organizers of *Quest* for permission to publish the worksheets in Appendix D, which help pupils over the first two stages of learning outlined above. Pupils have already had similarly well-constructed learning situations about reading other scales; the whole project is remarkable for its detailed programming of the stages involved in learning, for its individual learning approach and for its neat organization both of individual practical work and of records of progress.

Ability to draw

This may be the only skill a particular pupil shines at, in which case encourage him to feel successful in this, while not neglecting the other skills which he needs. To many pupils a picture or a diagram is the best way of showing comprehension, for them the best way of consolidating learning. Further elaboration on learning by drawing will be found on **page 78**.

Social and cultural

This section is intended to help the teacher become aware that if he or she generalizes from the experience of his or her own background, attitude and values on what kind of education his classes either need or appreciate, not only will the generalizations be invalid but the teaching can be miserable! Unfortunately this seems to be a warning which goes largely unheeded and the teacher usually has to learn it, by experience, in the worst possible place – in front of a class.

This is in part due to the dichotomy, given in the media and reinforced by college lecturers, of the working class and the middle class, which leads the new teacher into a simplistic approach. I, says the teacher, *am* working-class, so I know how working-class kids react. Unfortunately the so-called working class itself is composed of many subcultures; almost by definition the teacher will have come from one which values education generally, and respects teachers and the school as an institution. He then begins teaching where the prevailing values are that teachers are a boring interlude between work, money-earning and leisure. The general reaction to such cultural shock is 'These kids don't *want* to learn.'

The second aspect is that in some cultures positive attitudes to school work are provided by the home, and the school in its turn traditionally adds on to this solid basis. The idealists of comprehensivization therefore sadly provided the same sort of home-supporting education for all cultures, a knowledge supermarket where the new intake found that it had no acceptable social currency with which to buy the proffered wares. Realization of the various gaps in the home provision for educational skills has since resulted, for example, in the growth of the pre-school playgroup movement.

Thirdly, in the traditional home-supporting education, very often the pupils learned in spite of their teachers rather than because of them. If critical validation was not taught at school, it was caught at home; if there were no books, then encyclopedias were bought, or, in more modern times, subscriptions to science weeklies. If teachers were totally inefficient, parents would complain, but mostly, amused indulgence about a teacher's lack of control was deemed sufficient as long as the knowledge needed could be acquired at home rather than in the riotous atmosphere of the class.

The basic complaints made by the British Community Relations Commission about the treatment of immigrants in schools are that they are made to feel inferior, because different; taught to read books centred on a different culture; and made to conform to rules and norms which are foreign to them. Therefore, I am not proposing any special treatment for immigrant pupils, but merely ask that all teachers investigate very carefully the values of the cultures in the school in which he or she is teaching, whether Sikhs or Geordies, Chinese or Welsh. The Welsh and the Gujeratis, for example, have in this country more in common with each other than with many of the English; both cultures are intensely concerned with the survival of their language and consider that schools are an important factor in their battle.

So perhaps we should be grateful to immigrant cultures not only for their food, but for their use of colour in these somewhat beige-coloured isles, not only for the enlivening of British cricket but also for the awareness they have brought that not all the English are cast in one stereotyped mould. Unfortunately anthropologists seem to have neglected British native subcultures apart from tinkers and gipsies, so I can only give a brief indication of the continuum of differences which a teacher might find.

British indicators of the variety of native subcultures in Britain today

Childhood – in some cultures is a delight and to be prolonged, in others children attempt outward displays of adulthood long before it is achieved.

Signs of success – these vary from, for example, outward show of affluence to appreciation of job satisfaction.

Morality – some cultures are concerned with the intent of behaviour, others only with the consequences: it is OK to skive or cheat as long as you are not caught, property always belongs to someone else so why shouldn't you steal it. This applies to Local Government immorality as well as to boys scrumping apples!

Discipline – is obviously allied to ideas on morality; some cultures socialize their young into seeking repressive discipline; most cultures need help in learning self-discipline.

Curiosity and interest – some cultures socialize their young into acceptance of dogma, sometimes under a façade of reverence for learning, others promote exploratory behaviour. There is obvious relevance here for biology teaching, and also for the choice of leisure activities and the efficiency of the choices in producing desired relaxation.

Fate, the government, the council, everyone else are to blame for all unsolved problems – is this a cultural attitude or is it merely an indication of a lack of circuits in the brain to deal with problem solving? A lack of ability to deal with the abstract and predict consequences? Some educational projects have been concerned with this, notably the Moral Education Project of the Schools Council.[71]

Motivation

Habits of work, attitudes to traditional school methods and regulations, to repressive discipline and self-discipline vary with the subculture, and each subculture varies with the way in which it socializes the sexes.

Here for comment is the variety of motivation possible within a mixed ability class, depending on the subcultural mores of the pupils' parents.

By short-term aims, e.g. feeling of success and mastery; success in future job for early leavers; success in leisure; feeling of improved self-esteem. If short-term aims predominate among the subcultures in your school then work will need to be interesting and obviously relevant to real life, rather than academic and irrelevant.

By long-term aims, e.g. distant thunder of examinations; professional aspirations and job satisfaction. There is a willingness to suffer here and now for future benefits, accepting present tedium and hard work for a secure self-image in the future.

By technology: all cultures tend to socialize their boys into a love of technology; or is it genetically related to visual and spatial abilities? Stereotypes of girls do not include a love of metalwork or technical drawing, but, given a chance, girls enjoy both.

By people, especially children – is this social and cultural or does it depend on the strength of maternal and paternal drives? In most cultures this is sanctioned for girls but not all approve of boys being so motivated, although many are.

By authority and repressive discipline – gentle graduates never cease to be amazed that there are some pupils to whom repressive discipline is essential if work is to be done, although many deliberately seek repression only for the joy of flouting it! It is very difficult to re-socialize subcultures into *not* needing it; for example, the tawse is still used in Scottish schools.

By autonomy and self-discipline Many adolescents will respond well to the idea that those who behave as adults will be treated like adults, while those who persist in behaving like tiny tots will have corresponding treatment meted out to them. Some adolescents find it hard to distinguish between pandering to teachers' whims and so seeming to approve of school and all its defects, and behaving in a self-disciplined way. This may be due to the teacher's interpretation of what constitutes adult behaviour – for example, he may always favour conformist behaviour, whether it is self-disciplined or not.

But other adolescents will resent anything but the most subtle education in autonomy and self-discipline. They are socialized to think of the teacher as someone who must make them work or be labelled permissive, and teachers are advised to examine local attitudes to this before attempting motivation by this method.

Finally, various reports and surveys have been published on the merits and demerits of mixed ability teaching, but very few have emphasized the release from boredom of the most able which such methods have always provided, most of them worrying whether or not the able will be held back. For centuries, professional teachers have been using mixed ability methods; the records of our public schools provide incontrovertible evidence.

Assessment

Of the pupil, by the pupil   Remembering that some are motivated by short-term aims, it is essential for them, and pleasant for others, to organize methods whereby pupils self-check their own learning. This also helps pupils who like to see their work as small discrete lumps rather than stretching interminably to the far horizon. Self-checking lists and questions should, of course, pay attention to the wide variety of communication methods possible. They are also useful assessments of previous knowledge when used at the *beginning* of a piece of work. They can lead to a feeling of success and hence act as a motivation in themselves, as pupils fill in the gaps of their pre-tested knowledge.
Of effort as well as of absolute attainment   Persistence and determination in aspects of communications, skills and abilities at which pupils do not succeed instantly, need to be rewarded. But so, on the other hand, does the very bright pupil who always finishes first and is so often 'rewarded' by being allowed to begin his homework!
Of skills and application of knowledge rather than only recall of facts. When an oscilloscope in a selective school broke down, boys from the nearby non-selective school mended it, although they knew little of the theory behind what they were doing. It seems only fair that this kind of expertise be assessed and recorded to the credit of the pupil – but how to do it? In place of the usual practical science examination to assess skills (and there is grave doubt that it does so with any validity) several British schools have devised their own examination where such skills are assessed as part of the work done by the pupils during the course. Here is a set of objective ratings for a course work assessment designed by a group of London science teachers and reproduced by their kind permission:

Course work assessment – objective ratings

Knowledge, comprehension and interpretive powers	*Score*
– the ability to learn facts, terminology, generalisations and theories; to explain observations and to hypothesise.	
a) Demonstrates a high degree of recall and comprehension of observed phenomena; can hypothesise (concretely) intelligently.	5

Knowledge, comprehension and interpretative powers Score

b) i) Shows good recall and good understanding and hypothesising powers.
   or
   ii) Shows high degree of recall, but has some difficulty in understanding and hypothesising.   4
c) Shows a fair degree of recall and of comprehension.   3
d) Shows only fair recall and understanding even when helped.   2
e) Shows little ability to recall facts.   1

Critical ability

– the ability to make a reasoned judgement on the validity of a conclusion, argument or experimental design.

a) Demonstrates a high degree of critical ability, by being able to criticize logically and constructively. Demonstrates a high degree of open-mindedness and of creativity.   5
b) Shows a good ability to criticise experimental design constructively, but has more difficulty in the constructive criticism of arguments and conclusions.   4
c) Can criticise well – but only destructively. Can criticise constructively when helped.   3
d) Shows only a fair ability to criticise, with few constructive components – even when helped.   2
e) Shows little ability to criticise.   1

Accuracy

– the capability for:
   i) making precise, correct measurements using science apparatus
   ii) observations of natural phenomena.

a) Is capable of measurement with all kinds of scales and measures. Has excellent powers of observation.   5
b) Has good powers of observation and can use all forms of measures but often makes simple mistakes in reading scales.   4
c) Shows good powers of observation and can use all measuring instruments, but not very accurately. Mistakes are more fundamental (e.g. meniscus, parallax). Responds to help.   3
d) Shows fair observation powers but has more difficulty with all forms of measurement. Responds to help.   2
e) Shows poor observation powers. Has great difficulty even when helped.   1

Recording

– the ability to write down (and draw) observations and experimental details.

a) Shows an outstandingly high ability to make logical records of observations accurately, such that a complete record of the work done exists. Writes up work when absent from school.   5
b) Shows a good ability to make logical written records of experiments fairly accurately. Record of work is not always complete. Usually writes up work when absent.   4
c) Writes an account of an experiment but write-up is often illogical in order and observations are recorded haphazardly and incompletely. Does not discriminate in record between significant and insignificant observations.   3
d) Record is only fair. Needs a lot of encouragement and help – responds to this.   2
e) Record is very incomplete.   1

These are the categories which are to be assessed during the actual performance of the work in the laboratory by the pupils; each category carefully specifies the level of work given to each rating. In this examination some thirty pieces of work are assessed in this way over the two years of the course. Samples from the work of each school and class are submitted to an externally appointed moderator who compares and regulates the standards of the ratings given.

At 16-plus, the pupils take two written examinations which contribute 60 per cent of the final mark, the rest being provided by the course work assessment. To devise and implement such an assessment is a difficult, but a rewarding task.

Teachers should not confuse course work assessment with continuous assessment; eight pieces of course work assessed per term does not mean that Big Brother is always watching you!

Practical problems and operational effects of implementing such a course work assessment

Teachers' doubts

These are commonly expressed as:

Will this five-point scale work? It must be recognized that a five-point scale is intended to cover the whole of the range of ability, which may well not be present in any one class. Thus, one particular teacher may only be operating a three-point scale. It has been generally agreed that to increase the number of points on the scale makes precise allocation difficult and that the use of only some part of the scale does give comparability over the whole range.

Will I be able to control, teach and assess the class at the same time? This depends on the design of the assessment instrument. For example, a worksheet can be designed which tests accuracy of observation in the main, although obviously other skills and abilities will be needed for its completion. Where group work is being used, it is found that after a little practice it is possible to assess the contributions made by each member, acknowledging that each member of the group must be assessed at a different time and not when answering the same question.

Will I be objective in my assessment? When course work assessment is introduced a number of meetings for discussing methods and standards are essential. For this purpose many groups regard the first term of the operation of such methods as purely experimental and the results are not recorded officially. Similarly, when a new teacher comes into the scheme, his first term's assessment is regarded as part of his *own* assessment, and the pupils need to be given extra assessments in the terms following for the official record.

Effect of course work assessment on the teachers
After initial apprehensions and readjustments, teachers generally agree that course work assessment means a good deal more work and record making, but feel that the increased satisfaction found in teaching more than compensates for the increase in work.

Primarily the satisfaction rests on the effect of well-designed course work assessment on the pupils and the realization by the teacher that the pupils need to know how and why they are being assessed. Instead of a long-term assessment by an anonymous authority, pupils react to a simple short-term aim personally assessed by the teacher in the immediate present. They can be encouraged to assess themselves, one of the most important aspects of any adolescent's education, and one too often totally neglected by traditional secretive methods. Carrying out course work assessment during a lesson may include telling a pupil that he has fallen between scores 3 and 4 and encourage him to raise his achievement level. On the other hand pupils with an off day must be reassured that just because they have received a low attainment in this particular assessment, it does not mean that they cannot pick up higher scores on others. This also provides the pupil with some measure of protection against unfair assessment. It is impossible to argue (and win!) with a far-distant external examiner – but with the examiner in class, with a jury of your peers, and, most important of all, a non-secret assessment, unfairness should be minimal. Pupil involvement in assessment is an essential part of the process of making school-work and, in particular, the values attached to the behavioural objectives of science education, significant and relevant to them through experience rather than through exhortation.

Flexibility
Teachers are individuals and will show a range of variations in their judgements. Rather than attempt to standardize behaviour, teachers on the whole favour the use of an adjustment factor – assessments are compared, a standard is set up and each teacher given an adjustment factor which will

Mixed ability teaching

vary with the category being assessed. Such adjustment factors are re-allocated at meetings.

Thus the authority and individuality of teachers, which has previously been undermined by traditional examinations systems, can be restored, and most teachers feel that the hard work involved is well worthwhile, although many would welcome official help such as extra timetable periods for this work (time is always scarce for science teachers) and help with typing, duplicating and record keeping. One local authority has already surveyed such needs using time and motion study methods. From the point of view of time/cost/benefit analysis it is obviously a waste of expensive expertise for such work to be carried out by teachers.

Absence of the teacher

When a teacher involved in course work assessment is absent for a long period, the procedure for course work assessment is obviously at risk and previous provision must be made for this, possibly by the use of another teacher in the group with the caveat that the substituted teaching which the class have had will have, in most cases, an adverse effect on their work. This does not mean that the substitute teacher is necessarily worse than the absentee, but that the time he takes to establish communications and new work patterns will, of necessity, slow down progress.

Finally, get some ideas of other people's views of mixed ability teaching and ways of tackling it.[47, 59, 60, 101]

4 A variety of methods

Independent learning

During the later 1970s a trend towards a different approach to learning developed in school science laboratories. This was a move away from teacher-paced learning towards a more pupil-paced approach.[51, 91]

It coincided with the growth of mixed ability science classes and its parent was the programmed learning approach of a decade earlier.

During this time the staff in many school science departments, particularly in the larger comprehensives, got together in devising science courses based upon a variety of independent learning approaches. Most were initially based on an integrated science course in the lower secondary school. The results of some were eventually commercially published.[5, 61, 105]

Independent learning involves a wide spectrum of activities, from the use of worksheets during an hour's lesson to a completely individualized system, operating continuously for a school term or longer.

Don Reid (1973)[86] reported on the advantages and disadvantages of long-term independent learning for biology.

'Advantages

1 Abler pupils work faster, without any loss of knowledge gained.
2 Pupils take more responsibility for their work.
3 Teachers get to know their pupils better.

Disadvantages

1 The less able become increasingly bored and depressed.
2 It proved very difficult to organize an adequate supply of living material for long-term independent [learning]. Too many items were needed in small quantities at all times (e.g. fish, locusts, microbe cultures, etc.).'

Areas where forms of independent learning may be particularly effective are perhaps in the upper forms in the school. Currently there is a project funded by the Inner London Education Authority, 'Advanced Biology Alternative Learning',[1] developing materials for use in the sixth form. After all, an

A variety of methods

independent approach to study is one of the aims we might hope to develop at this level.

The Biological Science Curriculum Study have produced a series of units for independent study – their 'minicourse'.[82] These are based on audio-tutorial materials and are designed for college biology students.

Programmed learning

Programmed learning can play a role in independent learning schemes. The first job is to break the course into small steps. This is a very sound exercise for a newly qualified teacher and, in fact, an essential part of any good teaching. Any well-constructed worksheet or well-planned demonstration does this, and perhaps one of the best examples of this approach is Don Reid and Phil Booth's *Biology for the Individual*[87, 88] which developed from the Nuffield Resources for Learning project.

Programmed learning is based on the process of operant conditioning – a rat in a laboratory investigation learns when offered food as a reward and the student in the same way is led on by the spur of success. What then is success? How much depends upon initial motivation? On a course of programmed learning when students are first given a programme, especially when it is on a teaching machine, novelty is the spur, but this gradually wears off. Is it a reward when a printed page says 'Good – now turn to frame 99'? Is the intellectual delight in mastering a programme common to all subcultures? On the credit side it must be noted that even the least able pupil takes a delight in answering quizzes.

The essential idea of the programme is to split the information into frames. Then give one piece of information per frame, ask a question, reward the right answer – this is the bare bones of the idea. Cues and prompts are given to help in finding the correct solution, and there may be a review of the sequence of the end of the series of frames.[20] Establishment of facts and relationships may precede rules or a rule may come first and then be reinforced with several examples. Corollaries to the learned rule may then be made and reinforced by discrimination tests between it and the basic rule. This is obviously very relevant to mathematics – how relevant is it to biology?

Validation of programmes

The state of knowledge must be tested before the programme and then the increase of knowledge after the programme. Unfortunately the 'Hawthorne effect' is operative: performance improves in those who have attention given to them, notice taken of them – whether in time and motion studies in a factory, trial schools in curriculum reform projects or trial classes with a new programme. Hence some of the early spectacular results of programmed learning led fanatics to acclaim it as the educational Messiah! Like all other methods it has its place, which must be critically determined once the Hawthorne effect has worn off. We have yet to see how successful independent learning approaches will be. Certainly they have had some encouraging successes to date.

More usual methods of teaching biology

Even a strict programmed learning approach does have its value because it involves precisely sequencing information and then reinforcing it by asking the learner to apply it. However, most will feel that it is alien to the usual methods in biology teaching of proceeding by practical work, drawing and discussion. Familiarity, however, does breed contempt in this case, and too often new teachers begin teaching by talking about the topic, then searching, often desperately, for some practical work to do or something to draw, finally 'discussing' the work (which usually means teacher talking *at* the class again) and the pupils writing notes. Whether this is a learning situation or not is doubtful, and all new teachers are recommended in the first instance to try the independent learning approach in the Reid and Booth books.[89],[90] Using these books with a class is one of the best ways for the teacher to learn by experience what can be a good learning situation.

Now let us closely examine the pillars of biology-teaching wisdom.

Practical work

Teachers need to look critically at proposed practical work – is it merely recipe-following without comprehension, the biological equivalent of sand and water play in the nursery school? Or are the class really learning from it? A classic example of a waste of time and effort, with a high degree of anti-conservation, is the work recommended in a section of the first edition of Nuffield O-level biology which aimed at helping pupils to see fertilization as it actually occurred. The idea that sperm meets egg is easily understood from a quick scribble on the writing board, or a pair of simple models; ideas on relative size are appreciated by confronting the class with the idea that the whole of the world's population has been produced by a pail full of eggs fertilized by a thimble full of sperm. Nuffield O-level, however, recommended the removal of eggs and sperm from *Pomatoceros triqueter,* a charming marine worm, and observing their fertilization under a microscope. 'The first-years do enjoy it' said one teacher. But are the first-years capable of using a microscope with such sophistication that they can actually focus on the plane in which fertilization is occurring, let alone on the precise egg and sperm and the actual moment of penetration? Or are these hapless little worms merely being butchered to provide a Nuffield holiday? The project did, however,

A variety of methods

make a film[124] of the event and in the interests of all it is perhaps more beneficial to use this rather than the real thing.

Using a film rather than live animals used to be regarded as low-grade teaching, as it provided evidence at second hand rather than first hand; this is yet another example of applying a valid principle uncritically to all situations.

Luckily no one has ever investigated the amount of taxpayers' money wasted every year with that firm favourite of all biology practical lessons, 'food tests'. The basic aim here seems to be to use up all available test-tubes and to ruin at least one pupil's clothes with boiling, bumping Fehling's solution. Yes, of course, they love playing about with test-tubes and coloured messes! But do they

> learn anything about the nature of different foods? Most seem to leave school unable to differentiate between foods which are mainly carbohydrates or proteins; fats are obviously fats, except when in pastry. The ability to discriminate is more efficiently taught by means of the 'Diabetic Diet' game (see page 110) or by means of a circus where a large number of applications of the rules of what contains protein or carbohydrate can be experienced. Only questions such as 'Is there protein in Mars bars? in bread?' cannot be answered by simple observations.

> learn anything about the logical application of a scientific testing technique? Rarely, as most schools seem not to bother about the use of controls or the use of a positive test as a reference. (See Figure 11.) Consequently, many pupils will say that starch has turned the I/KI solution blue-black when a comparison with a positive blue-black would show that the colour observed is still brown!

So beware the use of practical 'to give them something to do' or 'to fill up the lesson' or even to promote the teacher's image of the 'white-coated science expert'. There *are* teachers to whom practical work is a genuine extension of their own expertise and an excellent learning situation for their pupils; others use it merely as a support for a sagging ego, while their pupils learn only bad habits and poor science. Teachers need first to decide what the pupil needs to learn and then select the appropriate method of helping them learn it.

Finally, self-discipline is essential in laboratory work. Teachers need to plan the arrangement of the work, microscopes, bottles, of reagent, etc. so that there are no traffic jams, no cause for the yell 'who's got the methylene blue?', no written work in the line of fire from water taps, no cases or bags for pupils to fall over, no long hair swinging into bunsen burners, and safety spectacles available and *worn* when necessary. But the responsibility for clearing away must rest firmly on the pupils, if only because they make less mess if they realize that they will have to clear it up!

1 Starch: using test tubes

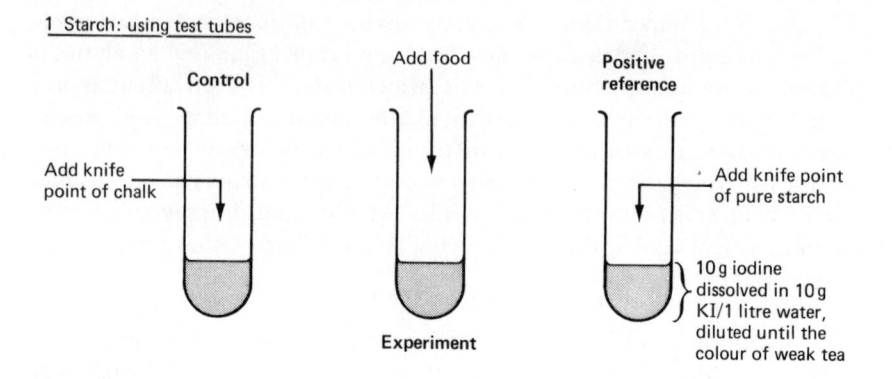

Control — Add knife point of chalk

Experiment — Add food

Positive reference — Add knife point of pure starch

10 g iodine dissolved in 10 g KI/1 litre water, diluted until the colour of weak tea

Using tiles

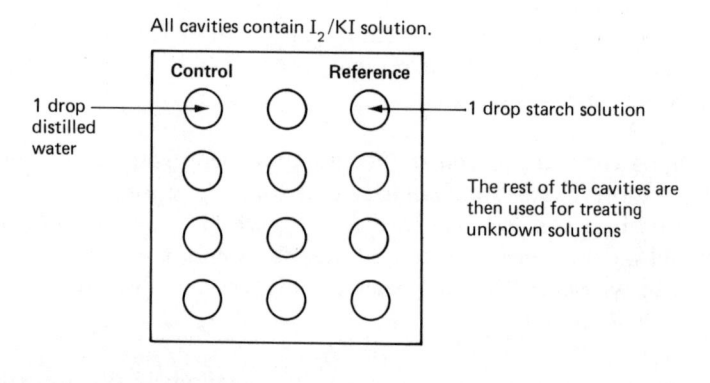

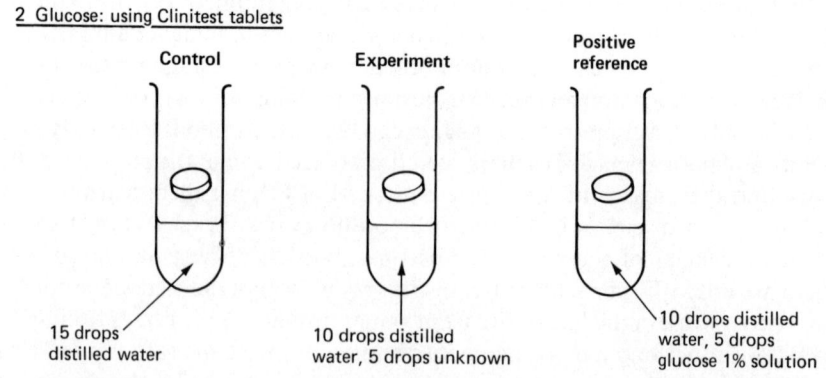

All cavities contain I_2/KI solution.

Control — 1 drop distilled water

Reference — 1 drop starch solution

The rest of the cavities are then used for treating unknown solutions

2 Glucose: using Clinitest tablets

Control — 15 drops distilled water

Experiment — 10 drops distilled water, 5 drops unknown

Positive reference — 10 drops distilled water, 5 drops glucose 1% solution

Figure 11

A variety of methods 71

'But Sir, how can you be absolutely certain it works, if we don't try it out?'

Teaching biology without a laboratory

What do you do if you have to use a lecture theatre or a classroom? The trouble with active science lessons is that pupils become conditioned to them – and when presented with a room that seems to rule out actually *doing* things, motivation sags. If the room is in fact used in this single period for giving back tests, teacher demonstration, lots of talk and no do, the pupils become conditioned to the fact that lessons in lecture theatres mean boredom, even if they are *not* cramped for space, which in itself stimulates the type of behaviour exemplified in shoving, poking in the back, etc. It was mentioned that similar behaviour is seen when a large class is grouped around a teacher demonstration in a laboratory, and also when given work such as in Nuffield Secondary Science on behaviour, theme 3.4, which demands no exciting apparatus but only boring-looking bits of paper.

Solution
(a) Reconditioning! Given that less cramped surroundings and the abolition of the single lesson are impossible, even if desirable, it might be put to the class that if they are to come to a lesson then they will not be made to work, it is their own responsibility to see that they do. Otherwise a teacher

can expect a battle every time (with ever louder shouts of 'settle down, 4B!'), achieving nothing but hoarseness in the throat of the teacher. Boys who still try to engage combat can be asked to bring a note from their parents saying that they have their parents' permission not to attend – 'But Sir, they wouldn't do that.' 'So, you have little option, Chris, but to attend and not stop or hinder those who wish to make use of the learning here offered.' Teacher expectations have been shown to influence pupils' behaviour – but the reverse is also true, so decline combat.

(b) Recondition to the fact that even this unpromising-looking environment can provide interest. Some simple practical work can be carried out such as work on perception, hand-held slide viewers such as 'Banta' viewers,[117] simple microscopes. It would certainly be a good spot to show films, if blackout is available.

(c) Recondition to the idea that bits of paper can be interesting – provide quiz-type worksheets, class questionnaires. A very simple technique again comes from Nuffield Secondary Science, theme 3.4 – the beginning of observation of human behaviour; each pupil has a piece of paper at his place on which is written 'What is on the other side?'. The teacher observes the reaction of each individual to this stimulus, and eventually discusses and lists them with the class: who ignored it as having nothing to do with him, who turned it over, found nothing and gave up, who turned to his neighbour and said 'But there's nothing on mine,' and who made a paper aeroplane out of it, set fire to it and aimed it at the teacher – a not-unheard-of occurrence.

(d) Use the environment for those lessons to which it is most suited, such as presentation of projects, results, or papers from pupils to pupils. If a pupil is occupying the demonstration bench, the interaction changes. Mass investigations, such as finding the range of frequencies which can be heard by the class, are possible. Groups of pupils can prepare demonstrations to give during this lesson – although the setting up of complex work may well involve an impossible time factor.

Learning by discussion

Class 'discussion' all too often means either a teacher talking to himself with pauses while waiting for answers which never come, or the teacher and five interested pupils holding a conversation while the rest of the class go to sleep. At the other end of the continuum there is verbal anarchy, modelled on parliament or on TV debates. If the class is already conditioned by years of experiencing such anarchy, it is hopeless for a new teacher to attempt to change it just by asking, and he or she usually descends to the irritated shout 'Put your *hands* up' 'I will *not* take shouted out answers!' and finally 'This class cannot discuss, I have to use worksheets'. This last is a sensible idea; the

A variety of methods

teacher can work up from well-regulated discussion in small groups gradually to a civilized discussion by a whole class; well, sometimes, anyway.

There is a case for the 'How to talk and how to listen' lesson, in the first week of the first year, giving some strong motivation for discussion, possibly during a session on laboratory safety (see the section on page 129). The ideas of the class on discussion should be actively sought, rather than imposing teachers' ideas; the pupils usually come up with some sound suggestions for regulating it which would appear totally fascist if made by the teacher. What matters is that the teacher works towards good discussion techniques and consciously praises success in this sphere – one often hears that 2B are brighter/nicer/nastier than 2C but rarely that 2C are so skilled at discussion that they can run one themselves.

Physical arrangement is difficult for discussion in most laboratories; in a lecture room at least pupils can turn round and discuss with those behind them, and can be in *eye contact* and *voice contact* with the teacher, although he cannot move around the groups. But in a laboratory too many teachers try to discuss from behind a demonstration bench, with the class fixed in rows stretching to infinity; every answer must be repeated for the benefit of the back row, and most teachers forget to do this some of the time, so the back row loses the thread and goes to sleep or rack and ruin. The only solution to a whole class discussion in a fixed laboratory is for the discussion leader, pupil or teacher, to stand in the middle of the room – far from the writingboard, so that a recorder must be appointed if needed. If possible the room should be rearranged so that bench tops form a square or a U shape, giving maximum eye contact and something to lean on and write on. Rings of chairs are favoured in some schools for discussion work, but they can provoke pools of silence and have no working surface. (See Figures 12–15.)

Prepared points of discussion should appear on worksheets or on the board; unprepared questions tend to be vague, too general, or leading, unless based on years of experience – and even then . . .! Unfortunately discussion often has to take place around a fixed oblong demonstration bench where the writingboard is obscured from view, and the pupils at each end are either out of the eye contact or dampening their notebooks in the sink.

Previous discussion by pupils is a useful beginning to a whole class discussion – pupils who are embarrassed when asked to speak to everyone can then deputize another member of the group, although it is unwise to allow them to remain silent forever. Such silent watchers may be afraid to make mistakes in public – so ensure by previous discussion that they can answer *one* question correctly in front of their class and gradually build up their confidence.

Discussion of results is a useful place to begin, with the class putting its results on the board – for example, in work on variation. But they can also be helped by wall displays where groups are responsible for explaining their work to others.

Physical arrangements for discussion work

A Recommended

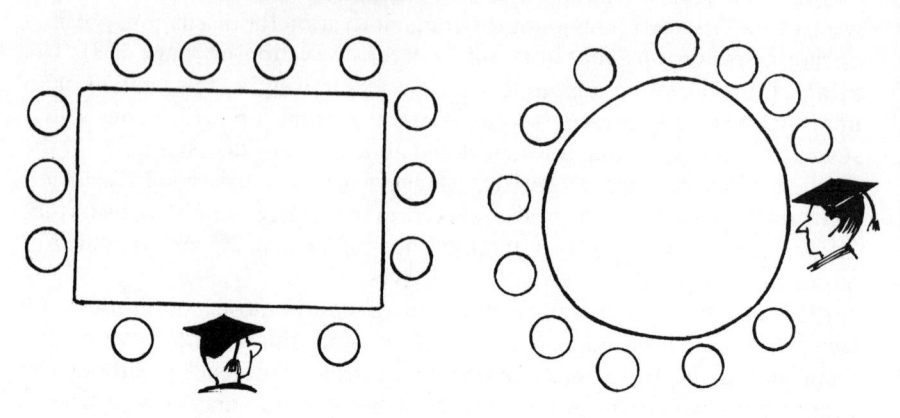

Some of these arrangements <u>are</u> possible in <u>some</u> laboratories.

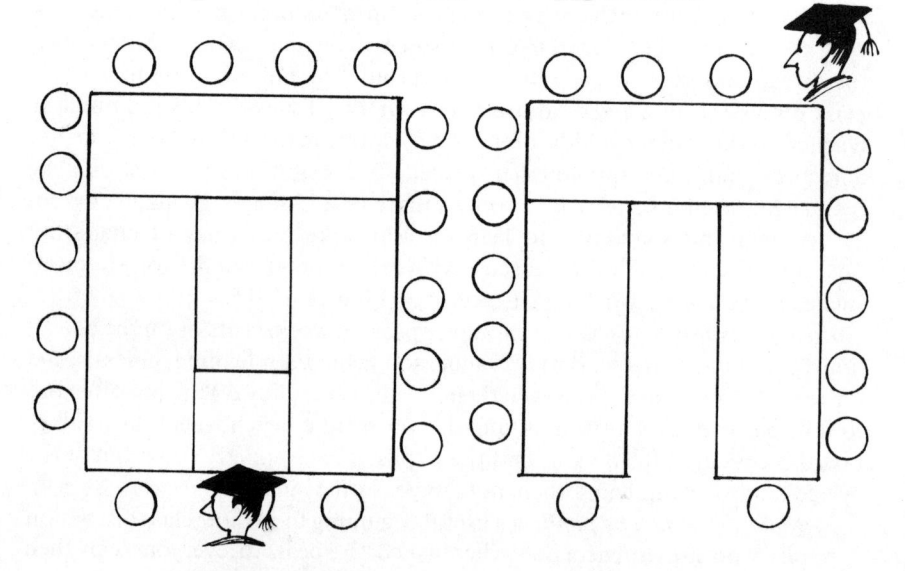

All the above arrangements:
1. Bring all the group within eye-contact — 12 is the best number to ensure maximum eye-contact and verbal involvement.
2. Provide a writing surface for all.
3. Avoid placing the teacher/leader/chairman in a dominant position.

Hypothesis — the demagogue/attention getter usually sits directly opposite the leader?

Figure 12

A variety of methods 75

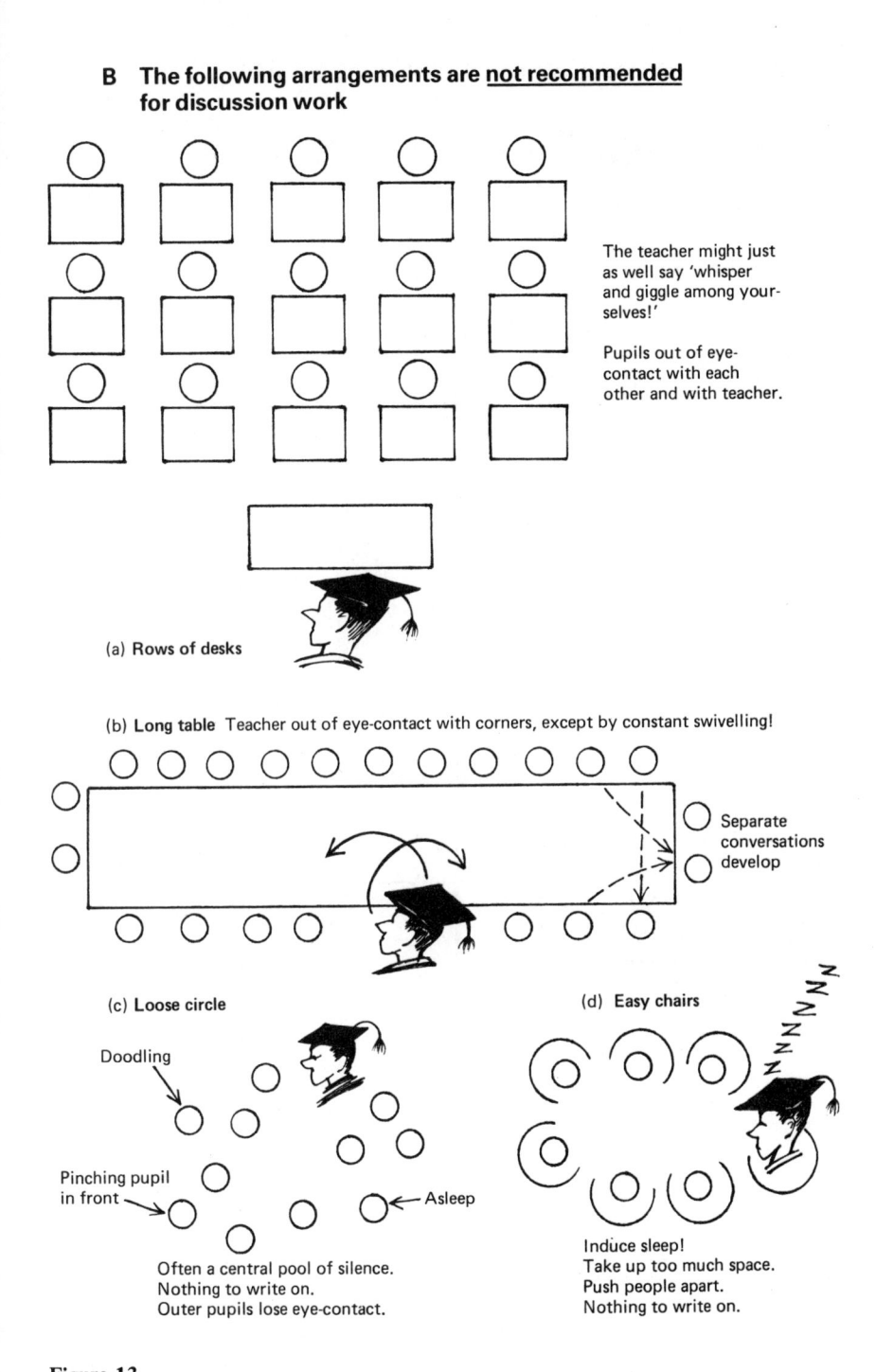

B The following arrangements are <u>not recommended</u> for discussion work

(a) **Rows of desks**

The teacher might just as well say 'whisper and giggle among yourselves!'

Pupils out of eye-contact with each other and with teacher.

(b) **Long table** Teacher out of eye-contact with corners, except by constant swivelling!

Separate conversations develop

(c) **Loose circle**

Doodling

Pinching pupil in front

Asleep

Often a central pool of silence.
Nothing to write on.
Outer pupils lose eye-contact.

(d) **Easy chairs**

Induce sleep!
Take up too much space.
Push people apart.
Nothing to write on.

Figure 13

C Discussion in a fixed bench laboratory

Teacher at board — has to repeat all answers from pupils so that back row can hear

Writing board

Pupil at board — for recording purposes

Demo bench

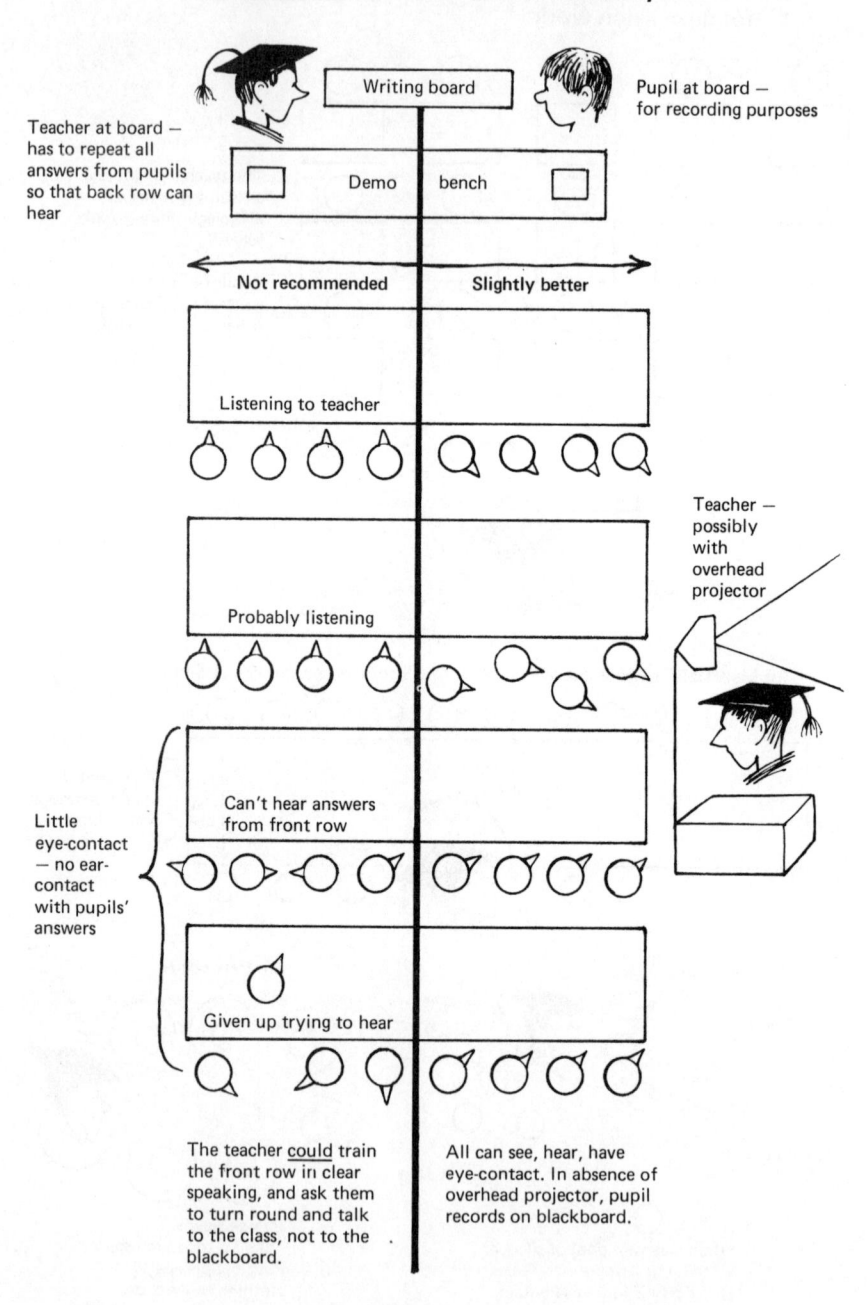

← Not recommended | Slightly better →

Listening to teacher

Probably listening

Teacher — possibly with overhead projector

Can't hear answers from front row

Little eye-contact — no ear-contact with pupils' answers

Given up trying to hear

The teacher <u>could</u> train the front row in clear speaking, and ask them to turn round and talk to the class, not to the blackboard.

All can see, hear, have eye-contact. In absence of overhead projector, pupil records on blackboard.

Figure 14

A variety of methods 77

D Discussion around a fixed demonstration bench

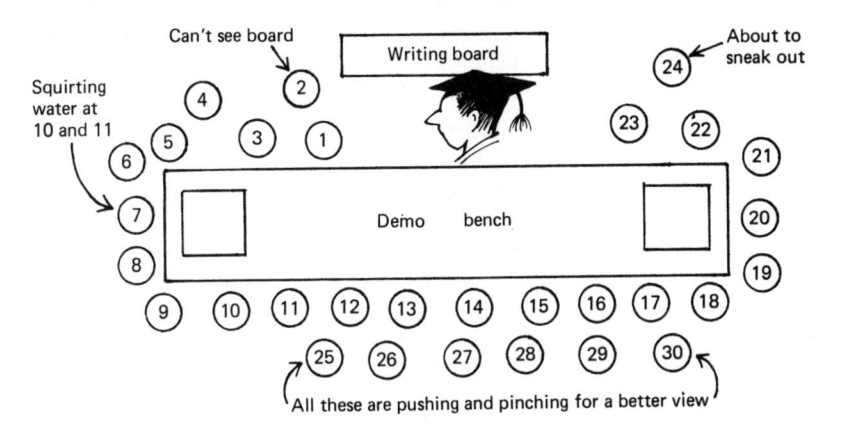

(a) An obvious case for small group work!

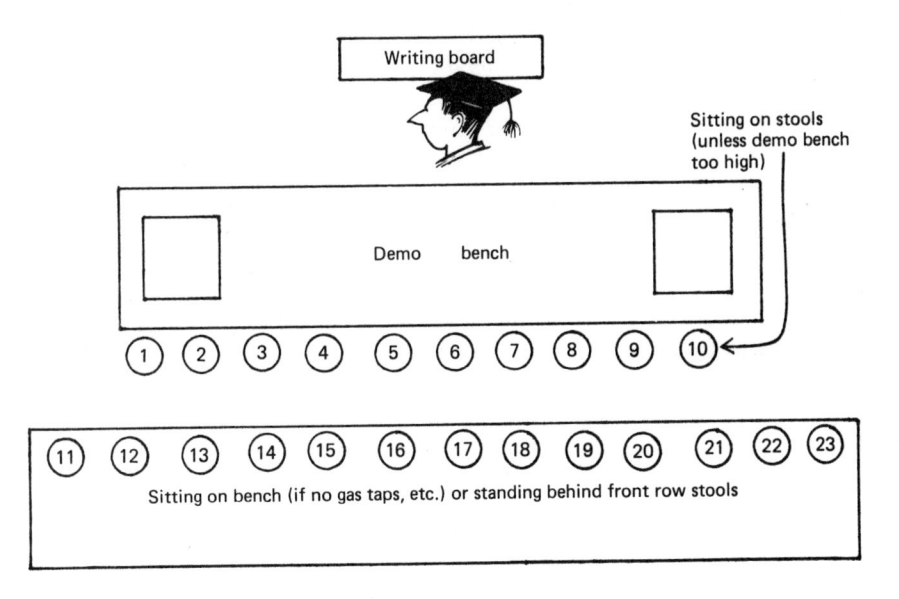

[Diagram showing students numbered 1-30 around a demo bench with writing board. Annotations: "Can't see board" (pointing to 2); "Squirting water at 10 and 11"; "About to sneak out" (pointing to 24); "All these are pushing and pinching for a better view" (pointing to 25-30)]

(b) When the whole class <u>must</u> see the board.
   Where could you put nos. 24–30?
   Stand on bench? Stand at sides of demo bench?
   Sit on stools behind and between 1–10?
   Depends on dimensions of benches, sites of gas taps, etc.

[Diagram showing students 1-10 in front row with note "Sitting on stools (unless demo bench too high)", and 11-23 behind with note "Sitting on bench (if no gas taps, etc.) or standing behind front row stools"]

Figure 15

A systematic approach and critical usage of discussion is, as in all other methods essential. Is your discussion really necessary? Do you need to lead it, or ask a pupil to lead it? Or should you try being a neutral chairman – dropping in pieces of advice or evidence from time to time? This is certainly a technique to try in discussions on behaviour and attitudes, but like all other methods should not be used inappropriately or to excess.

Suggest ways of silencing those who always want to hold the floor, including yourself. Read Abercrombie;[2] his ideas are particularly applicable to sixth form work, but the concepts are valuable for all age groups.

And with a silent sixth form as well as a riotous fourth form, you may have to concede that there are times when discussion is impossible with your class.

Learning by drawing

'I said to the first-years, "Draw a plant, any plant", and I got absolutely all kinds of drawings; tiny little fuzzy representations of an oak tree, enormous pop art, felt pen, jazz coloured pot plants and every degree of variation in between. What should I do about them?'

Darwin couldn't draw well, nor could Linnaeus, but most of today's teachers have been conditioned over the years to spend a great deal of time drawing; when they come to analyse the value per unit time of all this work they often divide it into the following categories:

(a) Looking at objects and then learning to reproduce a drawing useful to satisfy the examiners, who will accept it as evidence of his knowledge of the innards of *Spirogyra*, clearly labelled, i.e. *learning a visual code.*
(b) Looking down a microscope and trying to equate what one sees with the drawing in the histology textbook i.e. *verification.*
(c) Depending on one's artistic ability – either avoiding drawing at all costs, or enjoying the production of a really superb set of drawings of primate skulls, i.e. *aesthetic delight or innate aversion.*

Obviously (a) and (b) are valid with respect to the passing of necessary examinations – but the first form in question are some way away from this ultimate Gehenna; one needs perhaps to keep it in mind but also to examine some of the uses of drawing and some of the ways to spend time on it to greater value than merely copying textbook symbolism.

Training in perception

The first form are not aware that their eyes are receiving so many stimuli from the outside world that their brain rejects a large number of them, with the result that they need to learn to over-ride this selective perception if they are to observe things scientifically. A useful subject for observation is the human eye (Figure 16). If the first form are asked to look into each other's eyes and draw what they see, the results will be similar to those described by the new

A variety of methods 79

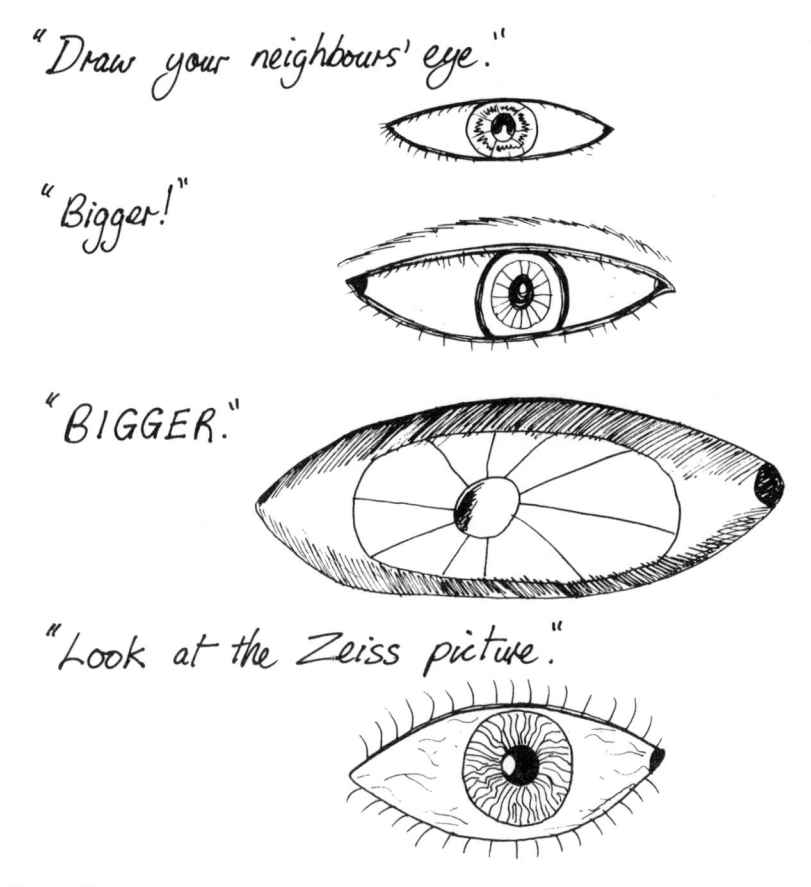

"Draw your neighbours' eye."

"Bigger!"

"BIGGER."

"Look at the Zeiss picture."

Figure 16

teacher who asked them to draw plants; some examples are shown here – drawn by 13-year-olds. Nevertheless all but the most disaffected will have observed something and if the teacher can gather on the blackboard a composite diagram of the most accurate observations, most of the class will realize their initial inadequacies and be ready to observe more accurately when the next object is given.

At a higher level of complexity, and particularly when dealing with history, perception needs training and testing too. A-level students who are quite familiar with textbook representations of animal and plant microscope sections are often completely at sea when given an unknown containing what should be recognizable structures. The Freeman and Bracegirdle, and Bracegirdle and Miles, series of biological atlases[15, 16, 40, 41, 42, 43] are useful here as they provide a photograph with a labelled drawing of it side by side. Teachers would do well to consolidate plant and animal anatomy, whether the studies are at the light microscope level or electron microscope level, to ensure that their students are capable of interpreting new work and recognize familiar structures in unfamiliar situations. A transverse section of the tail of a young rat or a section through the nose of a young mammal, for example, contain the whole range of tissues and include some usefully confusing tangential sections of hair follicles.

Training in symbolic representation
It is a pointless exercise to ask the first form to draw a plant, a fish or an insect without some training in rapid symbolic representation; they will spend a good deal of time involved in frustrating artistic attempts and the result will be neither scientifically nor aesthetically pleasing. Training in symbolic representation needs to be gradual and can follow these stages:

1 The teacher gives an incomplete diagram – the class observes the subject and adds the sections omitted by the teacher. More and more sections may be omitted as expertise grows.
2 Pupils observe the subject and attempt, with the teacher's help, a three-dimensional model, for example of a tooth, *Spirogyra,* a cell. Discussion of how to represent these models, either as a whole (*Spirogyra,* a red blood cell) or as a two-dimensional diagram (a tooth in section) may be done as group work and a composite diagram of the best ideas produced. Slicing through a plasticine model of a tooth helps to improve the understanding of the maze of tramlines in the textbook section of a tooth.
3 When classes are using microscopes for the first time, as when observing cells, they should relate what they see to a model (Figure 17). Without strategies like the one on the right, cells seem to be thought of either as wire netting or, singly, as poached eggs. Always, when working with microscopes, check that the pupils are actually looking at the subject under discussion. Many stained cheek cells have escaped the attention of the first-form pencil, which is drawing air bubbles, mucus strings, if not bits of that morning's scrambled egg.
4 Relating cell size to reality – if possible some kind of scale should be affixed to microscope drawings, but whether this is simply numerical, or visually represented (for example in relation to the head of a pin) depends on perception; perception of scale in microscope work needs to be consciously taught, and the Nuffield O-level chemistry film loop 'Measuring the very small'[133] is useful in this respect for above-average classes. (Figure 18)

Aesthetic education
Biology teachers are so often too busy to attend to this, although there are notable exceptions. Almost all laboratories contain biological objects which are worth looking at from an aesthetic point of view as well as a scientific one. It makes a good end-of-term lesson if classes are asked to choose their favourite things to form an exhibition for others to see. Those who enjoy

A variety of methods 81

From cell models to symbolic representation

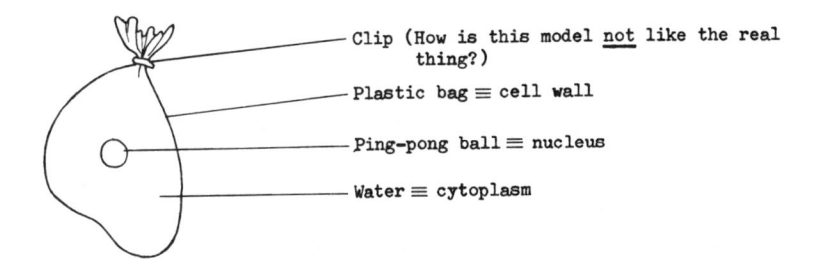

- Clip (How is this model <u>not</u> like the real thing?)
- Plastic bag ≡ cell wall
- Ping-pong ball ≡ nucleus
- Water ≡ cytoplasm

1. Fill a dozen small plastic bags as above.
2. Pile them into an aquarium tank (they leak!) - with all the clips <u>away</u> from the pupils.
3. Ask pupils to imagine what a slice of cells would look like, say if they cut a slice out of the pile of bags in the tank. Ask them to draw it.
4. Then push the cells up against one wall of the tank - do the drawings check with this view?

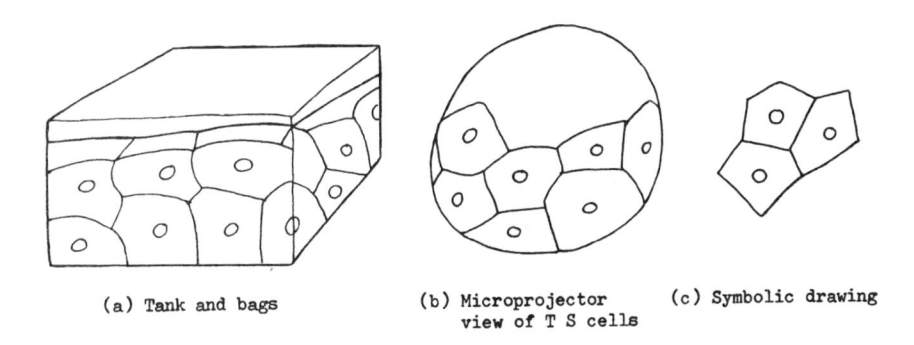

(a) Tank and bags   (b) Microprojector view of T S cells   (c) Symbolic drawing

5. Finally compare tank with a microprojector view, microscope view and consolidate the perception of cells as three-dimensional objects, usually viewed as slices.
6. Class or teacher produce three-dimensional models of specialized cells, e.g. plant cells, red blood corpuscles, etc.

Figure 17

Relating cell size to reality

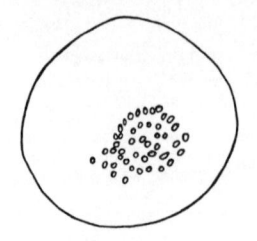

Too frequent style of drawing of first view down microscope!!

It is often better to ask the pupils to draw three views:

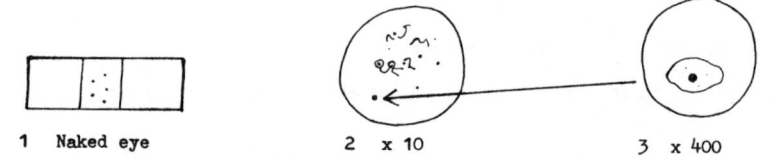

1 Naked eye   2 x 10   3 x 400

1 First a naked eye view.
2 Then a sketch map of the whole field of view.
3 Then indicate on the sketch map which area has been selected for higher magnification.

For example, in onion epidermis:

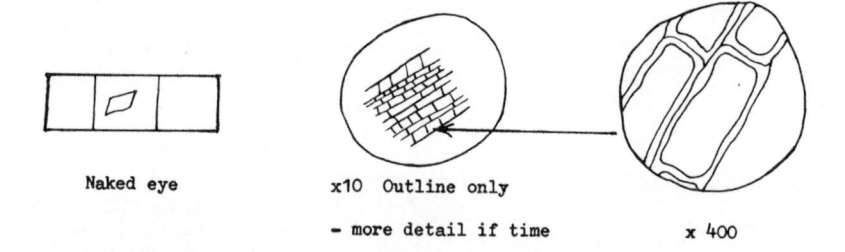

Naked eye   x10 Outline only   x 400
                 – more detail if time

<u>Teachers must decide</u> whether this exercise is, in fact, necessary for lower school pupils (it is essential for first year sixth). Would a model of the onion epidermis cell, together with a 35 mm slide or a microprojector, be better from a learning efficiency point of view?

Figure 18

A variety of methods 83

drawing should be given the opportunity to enjoy their skills in the biology laboratory – those who can't, can often turn out to be better at model making, if the teacher can invest money in plasticine, or acquire polystyrene packing which is easily worked with a hot wire or a heated knife. Almost all pupils enjoy violently-coloured felt pens – which can be a useful way of consolidating the learning of those tedious diagrams needed to convince examiners of the candidate's knowledge. Just because they are enjoying themselves, it does not mean that learning is being excluded from the activity.

Finally, a useful project for those depressed by their inability to draw is a little research into the drawings of Darwin and Linnaeus, or what they did instead; it could be argued that in overcoming their handicap in this field, they contributed more than they might otherwise have done to the pursuit of biological communication.

Project work

Too often this is merely fact-copying from books or cutting out pictures from magazines. Teachers should ask themselves what is the point of project work:
(a) Is it to motivate reading, given a series of questions, the answers to which need books?
(b) Is it to investigate some problem, using a practical technique, collect statistics, pictures, diagrams for presentation to the rest of the class? 'Growth' is an all-embracing topic which lends itself well to class projects.
(c) Is it to help a pupil widen his knowledge in a field of work not included in the syllabus, for example marine biology or bird and insect aerodynamics?
(d) Is it to encourage a pupil whose only ability is in drawing? In this case, the project should be in a form usable for wall display.

Nuffield A-level projects[108] are essentially an investigatory use of techniques and the booklet on projects produced for this scheme is well worth reading.

The variety of visual aids

Visual aids need time, trouble and technicians; luckily many schools have good audiovisual aid units, some have excellent resource centres and trained technical help. Some even have closed-circuit TV, useful for biologists, especially with microscope work; when one pupil's *Amoeba* is dividing, a quick use of CCTV will show it to the rest of the class – alas, there is no colour, although some schools do have external colour receivers and video tape recorders. Nevertheless it is unwise to base a lesson solely on visual aids, without something to fall back on in case of failure – even if the machines working, power cuts are not unknown.

Slides

There are a good many suppliers of excellent 35 mm slides on biological topics. These can be used with individual slide viewers, which are almost pupil-proof, but not entirely teacher-proof, as one does tend to forget that batteries run down rather than up. Used with a worksheet these can form part of a varied lesson; used with a synchronized sound system where the commentary is recorded around the slide itself, and one of the variety of devices which are available for automatic projection, they can form a more complex piece of independent work, giving evidence and aiding perception and observation.

Film-strips

Most teachers cut film-strips up into individual slides and mount them. A film-strip is often too inflexible and some teachers tend to slog through them, frame by weary frame, as if they had no choice in the matter. Banta 'Biostrips' are short film-strips well protected from greasy fingers and tearing. They are viewed in a small plastic lens holder, under which is a white reflector; this merely has to be pointed towards a window or light source. These strips come with worksheets, which are not always appropriate and tend to be didactic and full of 'Note!'. But they do provide information about the strip and teachers tend to use them and add their own questions on the writingboard.

Overhead projector transparencies

Again commercial productions tend to be didactic and expensive and most teachers make their own, building up complex diagrams by laying one transparency on top of another, using reprographic machinery to print their own transparent copies of, for example, original electron micrographs. Transparencies of such things as X-rays are doubly useful in that they can be explained to a group using the projector and then pinned on the wall or taped to a window for consolidation questions on a workcard. In work with older pupils on biochemistry it is useful to have an overview, for example of the glycolysis cycle, on the overhead projector, using a writingboard to explain further complexities and different aspects of reactions. The overview then can be pinned on the wall – it needs a long vertical space – for final notemaking.

Models

There are several available for the overhead projector, such as a particularly useful one showing the effect of different solid shapes on the flow of water.[137] All models have inherent dangers in that the pupils think they are reality rather than only half-truth. Expensive models of human anatomy lend a

A variety of methods

Osmosis model

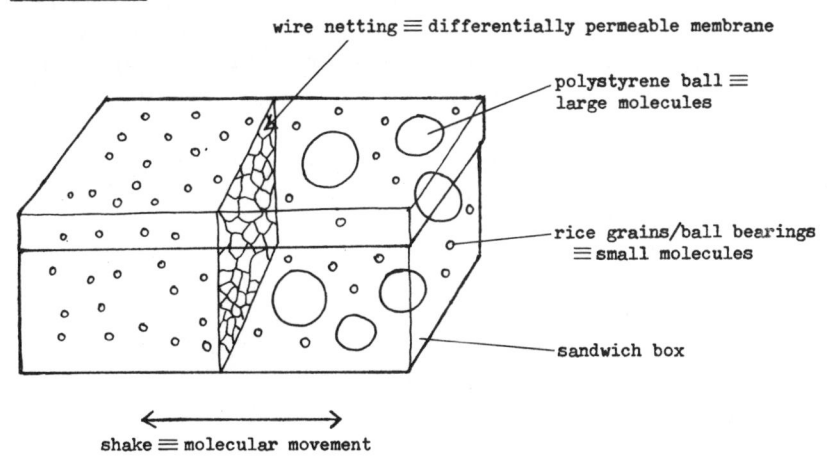

- wire netting ≡ differentially permeable membrane
- polystyrene ball ≡ large molecules
- rice grains/ball bearings ≡ small molecules
- sandwich box

shake ≡ molecular movement

NB: How is the model <u>unlike</u> the real thing?

Figure 19

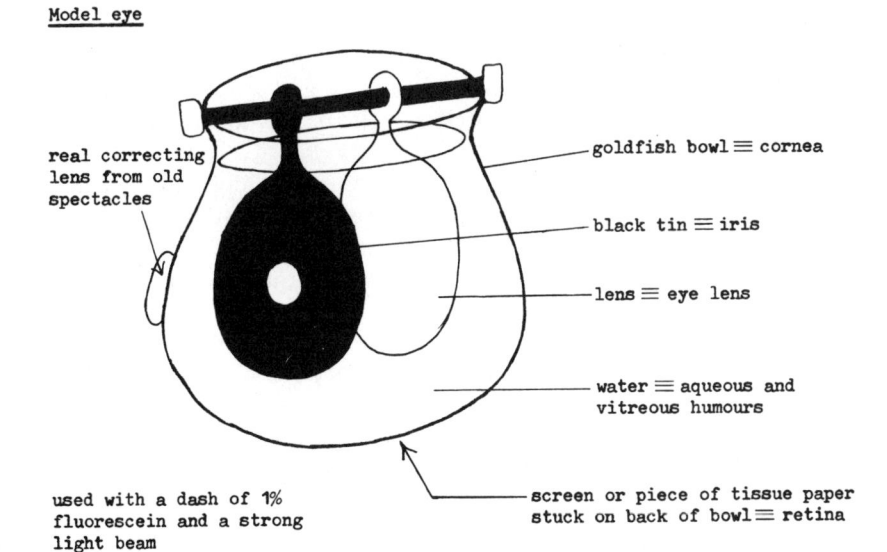

Model eye

- real correcting lens from old spectacles
- goldfish bowl ≡ cornea
- black tin ≡ iris
- lens ≡ eye lens
- water ≡ aqueous and vitreous humours
- screen or piece of tissue paper stuck on back of bowl ≡ retina

used with a dash of 1% fluorescein and a strong light beam

Figure 20

medical school atmosphere to the biology laboratory, but many are of rather limited use. Pupil-made models in plasticine, however, are valuable as consolidation and evaluation of work, such as on the anatomy of the tooth (see page 80) showing up clearly the perceptual problems of the pupils – and you can do a lot with beads in genetics.[54, 79, 128] A variety of objects can be used for modelling, and simple ideas are often the most effective – a transparent sandwich box divided by mesh large enough to allow small plastic beads through when shaken, but not larger polystyrene balls, gives a useful model of osmosis (Figure 19).

At the other end of the scale Bruce Mullin's model of the eye[74] is highly complex and most useful for showing the application of physical principles to a living structure (Figure 20).

16 mm film

Advantages: professionally made, usually good colour, good depth of focus, good camera usage.

Disadvantages:
1 Expensive to hire, often not available when needed. Difficult to view in advance unless you have access to a film library with viewing facilities.
2 Professional film makers are not professional educators; with relatively few exceptions, productions tend to be lectures on celluloid with enormous authority but little opportunity for involvement. There is sometimes a tendency to concentrate on what is usually exciting rather than what is important for teaching purposes.
3 Films are usually made for one type of audience and are unlikely to fit the range of variations in mixed ability classes with respect to pace, clarity and vocabulary.
4 Very often films are just too long – classes get mental indigestion watching them. However, the teacher has the power to switch off and take the film in small doses; although, all too often, it is a choice between letting a class see all the film, which is too expensive to book for more than one day, or to take just one part of it and teach it thoroughly.

With 16 mm film, therefore, the teacher must decide precisely what use is to be made of it – whether it is for consolidation, whether pupils should provide their own commentary, whether parts need to be shown twice, whether some parts need to be shown at all. A pre-test questionnaire or worksheet may often increase the value of the film. In addition schools which do not possess daylight screens for film teaching often tend to anaesthetize their classes, as those who design blackout for schools seem to take no account of its effect on the ventilating system!

A variety of methods

8 mm film loop/single concept films

Film loops came into vogue largely to overcome some of the deficiencies of the 16 mm film listed above. Film loops can be shown in daylight, some projectors have their own daylight screen, and the smaller cheaper variety need only an unlit, light coloured wall to project on, but the size of the picture you can get is very limited, even under the best conditions. Their flexibility in use is their chief advantage, and it is particularly useful that small groups can look at them during a lesson. The unsure technology of their projectors is their chief disadvantage.

Tape recorders

This section has not been entitled audio-visual aids, as tape recorders seem to have limited use in schools. They do not have the motivation of colour visuals and although some teachers find that the use of tapes is fitting for their own needs, they tend to use them only for a minor part of the work rather than as a central method. Exceptions noted in this book under the relevant sections include the use of tape recorders to help remedial readers with worksheets, or for recording projects and answers from pupils too embarrassed to face the class and talk. Pupils on the whole are fascinated by the sound of their own voice but have a limited attention span for the voices of others. Sound and vision together are more effective.

Synchronized tape/slide programmes

These have the same disadvantages as 16 mm sound films, their commentary makes them inflexible. Even so they have been proved to be more efficient than the lecture method in medical and nursing education.

Most teachers, however, if they use a sequence of slides, prefer to make their own to fit the needs of their class; commercial film strips are more often cut up and mounted as slides for this purpose. Exceptions could be quoted in the case of many of the TALC productions – Foundation for Teaching Aids at Low Cost (see Appendix E) – which are specifically designed for health care in particular countries.

Educational television

There is some evidence that the major effect of television in schools is to motivate. In the first flush of enthusiasm it was thought that the immediacy of television, with its ability to provide unique evidence and knowledge of experiments too dangerous or too expensive to perform in schools, would be of immense value. It was particularly useful that it was available in a form which did not need expensive film libraries with heavy maintenance charges. Many schools do indeed have video-cassette libraries, and many have colour television, often bought with private school funds rather than being officially provided.

Disenchantment came with the boredom rather than the motivation of the pupils when faced with patronizing presenters of programmes, interminably long sequences to produce one simple concept, the lack of knowledge of teachers' panels about television, and the lack of knowledge of professional television producers about education.

There is a move towards independent work using colour video-cassettes or cartridges, often divided into suitable lengths by teachers, and consolidated by worksheets, project work and problems. But this will depend on the material available – so much of it is confused and full of biological half-truths if not downright inaccuracies. Even the expensively budgeted science programmes for adults suffer from poor scientific editing so that, for example, wild and fascinating hypotheses are presented as facts – this may be sound journalistic technique but is hardly to be considered as good science education. What point in motivating and arousing interest and curiosity if the pupil then has to unlearn what has been put into his mind?

Some teachers set viewing of TV programmes as homework, and ask for a brief report on, for example, Cousteau or the 'Man Alive' and 'Horizon' programmes. Programmes for general viewing tend to have bigger budgets than Schools TV productions, and often the quality of visual evidence is higher. Some pupils have been able to produce large useful files of their viewing reports, and the discussion of last night's TV, which occurs anyway, is thus given reinforcement and consolidation and made into a useful and pleasant way of learning.

The use of visual aids

These have a variety of uses all of which need to be assessed critically and used systematically; make your choices from the following list:

Teacher's rest

The teacher orders a film, which comes at an irrelevant time in the class work, runs it through without comment or relating it to anything, then complains that the class is bored. There are schools who order films *en masse* for use at the end of term, perhaps to give the staff time for marking. Once a class is conditioned to a bored approach to film and educational TV, it is hard to change their ways.

A variety of methods

Active involvement of pupils

Yes, of course it means much more work for the teacher before the lesson, but fewer problems during it. The teacher may find it difficult to view a film before using it, and the word of friends as to its excellence cannot always be trusted. Nevertheless, if a teacher manages to work out why this particular film is needed, and how he is going to use it to involve the class actively in assessing the evidence it provides, a good learning situation can result and the use of the word 'Note', (generally equivalent to a lack of thought as to why they should or even precisely what) does not prevail. Pupils do get very tired of 'Note' to which the eventual reaction is 'Why should I?' – for which logical question there is often not much of an answer!

Total involvement of the pupils

Pupils can make their own filmed records of Zoo visits, field work and of experimental work in the classroom.

As an introduction, or as the first impact of new work

The private life of the head louse does not instantly motivate curiosity, even if introduced by means of live specimens, much less when introduced by drawings and talk. While there are some beautiful 35 mm slides[131] on the subject, there is also a 16 mm colour film,[138] which manages to make the beast appear positively aesthetic as well as scientifically fascinating! That which film and video do well, they do very well indeed.

Long-term experiments and processes

Whether made professionally or by the teacher and pupils, records of long-term experiments are often more conveniently and efficiently used when on film. It is important the pupils do not think that *all* experimental science can be performed within a double period. It is also important that the patience and perseverance of scientists, whether selectively breeding hens[135] or roses[120] or progeny-testing cattle,[134] are seen to be an essential part of the process. In addition, time-lapse photography, whether of growing plants, scudding clouds or human growth, adds a dimension to perception and learning impossible to achieve in any other way.

Visual models as aids to perception and concept formation

Whether the models used are three-dimensional as previously described, or on animated film, the dangers are the same. Films on sex education are notorious in this respect and girls have left a lesson fully believing that they have white tacking stitches down their Fallopian tubes or that their uteri are held together by something more akin to the bootlace on a skating boot than to human flesh! It is best, if this sort of film has to be used, to use the visual models as a 'spot how many things are wrong' exercise.

On the other hand, when a film is well made an average class may need perceptual education before they can use it. The diagrams in the film loop *The cleaning mechanism of the lungs*[122] are immediately clear to teachers with an education in the interpretation of histological diagrams behind them – but the classes they teach may need some previous model work before they can appreciate the film's sections through the mucous glands, the cilia and the bronchi.

Two Nuffield Secondary Science film loops, *Geological time scale* parts 1 and 2,[127] allow the formation of the concept of the length of time in geological terms; many visual attempts have been made at communicating this relationship, but these films do manage to squeeze 4800 million years into eight minutes and provide a valuable overview, possibly the nearest that a human being with limited perceptual powers can come to appreciating such a vast expanse of time.

Consolidation

Some films and loops intended for first impact work with those of high ability, are, in fact, excellent consolidations of work which must, for others, be taken in shorter steps and include learning by experience rather than by verbal treatment. If you look back to the lesson analysis on ventilation (see p.24), the film loop *Breathing*[119] is used there to bring together a varied pattern of experimental and demonstration work; other loops in the same series have the same quality. When used as first impact with average or mixed ability classes, they assume that too much can be taught by short verbal descriptions in too short a time, but when used as consolidation they are excellent. The short version of the film *Windows of the soul*[139] from Fact and Faith films could have been made for consolidation work on the senses and perception, and includes experimental work impossible to set up in the laboratory.

Demonstrating techniques

It must be admitted that many film loops demonstrating techniques are poor, others are out of date and the majority use a different technique to the one the teacher wants to use! The alternative is to make your own, but this has its pitfalls and takes a great deal of time and effort. Instead of film, though, a series of 35 mm slides with a synchronized tape commentary are cheaper and easier to produce and can be as valuable.

A variety of methods

Emotional education

The dissection of an eye can produce emotional shock in some pupils as they simultaneously are revolted by the sight of blood, the bouncy feel of the sclera when the eye is bisected and the jelly of the vitreous humour. A colour film of the eye brings the blood shock first, divorced from the touch shocks. Since blood looks so much more red on the film, when the pupils see the actual eye they comment that it's not so red in real life, become less hysterical and so more calm to receive the shocks which come when they actually touch the eye itself.

With birth films – blood looks very red, and babies look very blue. Again the separation of emotional shocks due to the colour and to the mechanics of the operation is advised; use an animated film first to show the process, then proceed to the real thing from the mother's point of view (after all doctors and midwives are paid to look after the other end) and, if it is really necessary, then show the real thing from the medical point of view. *The first days of life* (Boulton Hawker, from NAVAL)[125] does just this, putting the medical view at the end of the film so that the teacher may or may not show it, depending on the maturity of the class, while all can enjoy the birth of a baby from the mother's viewpoint.

Finally, don't forget the simpler visual aids such as felt pens and lots of paper.

How to use the writingboard

There was once a student who asked for a lecture on how to fill up a register; this is impossible as registers vary so much, and so do writingboards.

Some writingboards are also impossible – the chalk skates over a greasy surface, or the thing is so pitted with holes and scratches that your immaculate script will appear chalk white against a ghostly background only if you can press on really hard. Architects have thoughtfully arranged lights so that they reflect precisely off the writingboard into the eyes of the class. Then there are whiteboards, on which someone has always written with a coloured felt pen which is not of the kind recommended for use with that particular surface, and whose ink defies the efforts of the solvent provided. So whiteboards are often pink or purple at best, and indelibly marked with irrelevancies at worst.

Assuming, however, that you have a splendid, electrically operated, vast expanse of glass writingboard, how do you use it? Badly as far as I am concerned; even after twenty-five years of practice I cannot write either well or in straight lines, and hence prefer an overhead projector. New teachers are advised to practise writing on the boards they need to use, making sure that the chalk actually works (if it is there), that the erasers work, and that the lights help rather than hinder. Also, inspect your work from the back of the

'Used correctly, Helen, you will find the overhead projector an invaluable teaching aid.'

class – you may need to write larger. It is considered good manners to leave all apparatus in a state of readiness for the next person to use, and writing-boards are no exception, unless one of your colleagues has produced a magnificent diagram in ten different colours with 'Please leave' written on it.

Here again, I consider this a waste of time, probably because drawing was never my strong point! Drawing on the writingboard anything even as complex as a circle always caused me trouble. If a complex and beautiful diagram is to be done, I would prefer to do it once and for all on a set of overhead transparency overlays.

Teachers often use the writingboard as a visual twitch, writing down the odd word, scribbling the odd diagram, which, all too regularly, the class then proceeds to take down in their notes and confusion results. It is often necessary to use a writingboard in this way, but with the caution to the class that instructions for note-making will follow. When instructions are given, it is obvious that they should be clear and concise, whether for practical work or for a summary of theory, and for this layout and clarity demand as much thought as for a worksheet.

Teachers often have ambivalent attitudes to writingboards – on the one hand they seem to think that absolutely everything must be put on a worksheet even when the writingboard would serve the purpose well, while on the other they seem to consider that board work has not the authority of a worksheet, which, if it is disorganized and off the cuff, it certainly does not.

The critical use of the writingboard, one of the earliest visual aids, is still important.

Why not try –

A circus

Appendix B is a good example of the circus approach – the basic idea, the basic question asked is *'What is normal?'* A variety of measurements are asked for from the pupils, not merely static ones of height, weight, etc. but also operational measurements, such as 'How much can a hand pick up?' 'How far can pupils jump up?' In addition, two simple behavioural estimates are asked for – showing the variation in time taken for English homework and mathematics homework. The principle of variation along a continuum rather than the use of so-called norms can then be established by the drawing together of results given on the teacher's note sheet, and by the use of part 2 of the same worksheet, which also brings in the application of correlation.

A circus using experience to learn and apply the basic idea of energy transfer is found in Nuffield Secondary Science, theme 4.[78] Here pupils trace energy transfers through chemical, biological and physical systems, asking the simple question 'Where is the energy now?' while they blow windmills, operate steam engines lifting loads, watch sound move polystyrene balls, watch musical notes displayed on an oscilloscope, and eventually track the energy transfers from the sun to their breakfast cornflakes to the light they are producing from their hand-turned dynamo.

The main point of a circus is that one simple idea is applied in a variety of situations; it is not necessary for the pupils to go into complex verbal explanations of phenomena such as magnetism, momentum, inertia, etc. even if this were possible. You can then avoid the usual verbal comprehension testing, the learning of inappropriate definitions, and assess whether the pupils can apply the idea; for example, whether they can discriminate between instances and non-instances of the phenomena of osmosis. For osmosis, a variety of examples and non-examples can be shown, asking the questions 'Where is the differentially permeable membrane?', 'Is it dead?' (pig's bladder dried, Visking tubing), 'Is it alive?', 'If so, how does it act when killed?' (potato experiment, etc.). This is obviously not quite such a simple idea as energy transfer or variation along a continuum, as there are several factors to consider, several different questions which must be asked. And, of course, when variation and energy transfer are treated at a higher level, further aspects and complexities must be introduced. Circus treatment therefore establishes a basic simple frame of reference on which further complexities can be built.

Obviously circuses could be entirely verbal, entirely visual, entirely experimental, entirely based on experiencing the basic idea; one could call a physics problem book a circus encyclopedia – the basic principle is one of application. As many who have suffered physics problems will know, the ability to apply and slot numbers into the correct places in formulae does not necessarily indicate understanding or comprehension! So a circus in the operational sense, in common usage today, needs to provide a variety of different types of communication and application of the basic idea.

So what have concepts to do with circuses? Educational jargon uses the word synonymously with basic ideas, principles, generalizations, underlying theories. Markle and Tieman (1970)[65] give the following definition of a concept:

'A class or category, all the members of which share a critical combination of properties not shared by any other class.'

They give as the criteria for assessment of the learning of a concept:
(a) Being able to discriminate instances and non-instances.
(b) That being able to reproduce a written or verbal definition is *not* a criterion – thereby discriminating a non-instance!

In addition they indicate that *disjunctive* concepts form multifactorial situations. For example, there is a concept of squirrelness. All members of this class or category share a critical combination of properties, but a 'basic idea' of squirrels can range from an aesthetic appreciation to a hypothesis about where and why they put their nuts – a very disjunctive set of concepts. One can have a simple concept of respiration as the release of energy in biological systems, but when one needs to consider all the support systems of this mechanism, the gas exchange, the source of the energy supply, the systems involved in using and transporting that energy, etc. it is obvious that a variety of different concepts are needed, some of which are very disjunctive, and so we get complexity and multifactorial situations.

You don't only use the circus for teaching simple concepts to younger pupils. When teaching about hormones, the concept of targets needs to be consciously applied, and with it the concept of variation; for example female sex hormones do not only act on the reproductive organs at certain times of the month, Joan of Arc is said to have cried tears of blood!

In spite of Markle and Tieman, therefore, there is great difficulty in discriminating instances and non-instances of concepts! There is obviously a continuum of concepts, complex concepts, related concepts and disjunctive concepts. There are no prizes for arguing with the following lists, only that they help provide ideas for circus treatment, i.e. helping pupils to apply the concept in a variety of ways, demonstrating it in a variety of situations. And concepts can also be useful in the organization of knowledge – does one

A variety of methods

Table 4 *Different levels of concepts*

Simple	Complex	All-embracing
Of a leaf, a seed, a fruit	All on the left	**Energy** – work, field, wave, reflection, refraction, absorption, equilibrium, interconvertibility
Of plant families – even unto *Rosaceae* and *Scrophulariaceae*	Respiration	
	Evolution – includes gene pools, adaptation, variation, survival values, etc.	
Variation	Interdependence	**Structure** – relationships, force, particle, change of state, pressure, chemical change
Adaptation		
Steady state		
Energy transfer		**Chance** – randomness, pattern, disorder, uncertainty, probability
		Life – metabolism, inheritance, adaptation, diversity, interdependence

organize the sixth form's hormones work on lists of hormone actions only? Does one use the concept of target organs in the selection of comprehension work which would assess whether it has been learned and applied?

In biology, a mixture of disciplines, there appears to be no hierarchy of concepts, but an increase in their complexity of usage, which is wild and free and depends on the teacher rather than the accepted definitions. If the above list changes a teacher's paradigm of teaching that is all it can do – discriminating instances and non-instances is fruitless!

When is a circus not a circus?
One non-static class moving around a room from work-station to workstation, does not make a circus. Nor is it a circus when it is group work. Group practical work is very useful when you haven't enough apparatus for everyone, such as in the case of expensive stuff like sphygmomanometers, or stopwatches. So, for example, one group can do blood pressures, one can time pulse rates, one can use stethoscopes, one can do ventilation rates and one vital capacity. But this rebounds a bit on the teacher's classroom organization, as everyone finishes at different times, there are queues for apparatus, and so a judicious use of film loops, consolidatory worksheets or static demonstrations must also be organized. There is rarely a good reason for the whole class to be doing precisely the same experimental work simultaneously, unless of course you favour the band-master approach, conducting them through each movement, getting more and more allegro as bell-time approaches. Even when, for histogram purposes, the whole class needs to measure its height,

'On the down beat, cut with scalpel back from the first gill slit . . . '

weight, pulse, etc. this can be and almost always is done as group work, and the results recorded on the writingboard or an overhead projector transparency for class discussion.

Circuses can be a menace organizationally. If the apparatus they require can be neatly put into stacking boxes and then laid out around the laboratory for each lesson, this is fine. But too often, in over-used laboratories with little storage space and less cash, this is impossible – nor is it possible to leave the material about for a week so that all classes can use it.

On the whole, circuses do not pile up into queues of people waiting for apparatus – the experiences do not have to be followed sequentially, some take longer than others, perhaps because they require more complex thinking rather than complex techniques. What does happen is that particular pieces of apparatus hold more interest than others, and collect a happy crowd.

Worksheets

A multitude of variables are involved when one individual personality comes to terms with his own peculiar ways of teaching; some teachers never use worksheets, some use nothing else.

A variety of methods 97

'I've tried them – they're no good' is a common cry, even from teachers who have been on a course on how to write them. It's almost the same as saying 'Food, I've tried it and it was no good.' A systematic approach to evaluating the different varieties of food is provided by the taste buds and/or the weighing scales which will help the individual to decide what is suitable for various situations and needs; similarly with worksheets, only here the systematic approach needs evaluation via the teacher's perception of a particular learning situation. Some teachers use worksheets like lifebuoys, happy spirits who scrawl uneconomically over the school's limited paper supply and leave duplicating machines awash with dye – here, as in most situations in educa tion, the middle way should be trod. The following list of uses and abuses of worksheets is intended for critical appraisal.

Why?
To avoid repetition of routine instructions and questions. If there are only a few instructions, put them on the writingboard. If the instructions are for the use of a piece of apparatus or the investigation of a photograph, use a workcard, with questions and some help as to what the pupil should record.
To give detailed steps of complex techniques and investigations – these are easily referred to and hence lighten the teacher's workload in class.
To help in structuring the form of record, be they numerical records; diagrams and graph work; the record and consolidation of logical stages in learning.
To replace or organize written notes. Copying of notes into neat books is a less productive way of reinforcing work than giving an essay question leading out from the work, a question involving a practical application or a straight test question. If the class all have the same basic record on their worksheets, this can be done, and improves learning as well as cutting down on dull repetitive marking.
Used as a piece of independent learning with instruction for the inclusion of film loops, models, experiments, required reading, dissection, all in one integrated learning situation, and composing its record and consolidation.
Worksheets help an absent pupil to catch up and help the teacher to keep an accurate record of what was actually learned in a situation.

Why not?
Where the class is remedial, both reading and comprehension levels make the use of worksheets impossible, except when they are almost entirely visual.
Too many teachers have a tendency to write young textbooks. Too many words typed closely are a disincentive to anyone required to read them – as any student knows. Therefore:
Layout is important – study any well laid out advertisement, and remember that today's pupils are more sophisticated in these matters than their teachers were at the same age.

Priorities and safety precautions must be underlined or made to stand out with bold type or spacing.

Flow diagrams can replace wordy descriptions of techniques.

Paper is scarce and expensive. Instructions can be put on workcards and then covered with plastic film. Diagrams and photographs for observation are best used on workcards, unless they are needed for purposes of the pupil's record of work done.

Worksheets therefore are only absolutely essential when you need to train recording and comprehension. However scarce the paper, never economize on space; having decided that a worksheet is essential, to crowd it makes it visually unappetizing, and the spaces for recording answers must be large enough for pupils with a lack of muscular coordination.

Worksheets encourage teachers to think that, having provided one, they can then do nothing for the rest of the lesson except say 'Read the sheet!' No worksheet is ever perfect; you may also be using one written by another teacher whose teaching method is very different from your own. Most worksheets need some introduction and most need consolidation. To see a class mechanically filling in worksheets is as disheartening as seeing them recipe-following in practical work.

Using the worksheet

Comprehension is not automatically ensured by presenting sheets. Complex sheets may need to be given out for *previous preparation* – it is a waste of good laboratory time to spend it on reading. Previous preparation may include questions on what the class is actually going to do in the laboratory and even an element of experimental design and error-detection.

With the least able, some or all of the sheet may have to be read with them before work begins, together with actual handling of the apparatus rather than relying on purely verbal methods. Some questions may be left undiscussed for fast finishers. With less able pupils, again it is preferable to produce the work in small doses with, in addition, constant roundups and periodical reinforcement.

No teacher ever writes a worksheet which is perfect for every member of the class, but he is thus given time to cope with individual variation in learning.

Worksheets result in the lack of practice in verbal communication. For the less able pupil in particular, the ability to express new facts verbally (if not new abstract concepts), using and fully understanding new vocabulary is an *essential* part of the work. Therefore:

(a) Worksheets can be structured so that *group discussion* of the most full and satisfactory answers to the questions is promoted. This could be either in the course of the work, or, more often, through consolidation questions or

reinforcement used after the first impact and made on a communal basis. Pupils learn well from other pupils and from teaching other pupils – attention-demanders are often useful group tutors, benefiting themselves and the teacher as well!
(b) Particularly in groups whose written English is poor, such group discussion may lead to an *agreed common record,* and groups could then make a decision, as a class, on which group had the fullest answer. These agreed answers may then be duplicated with the name of the major contributor added; thus each pupil has a small amount of careful written work to do and a rather higher motivation to contribute to the class work than teacher's approval or a mark in a book.

Worksheets result in a lack of written work of the sort which gives practice in expression and synthesis. Let's face it, so do dictated notes and copying from the writingboard. Worksheets are an aid to less literate pupils, but the aim must always be to raise literacy levels. A progression may be used:
(a) filling in one-word answers – this at least implies the ability to read and understand the question (or to copy from the kindly child next to you);
(b) writing short sentences;
(c) answering questions necessitating a paragraph of short sentences;
(d) answering questions necessitating complex sentence formation, such as when a hypothesis is asked for with supporting evidence. Such sophisticated use, both of thought and sentence construction, may not be reached in less able groups until 16-plus.

Biological mathematics is an appropriate symbolic language which many need to be helped to understand and be able to express themselves by its use. A similar progression can be used:
(a) filling in numbers passively and using a simple formula as a 'black box';
(b) understanding the formulae they are using;
(c) constructing their own record sheets for easy and efficient use.

Similarly, in graph work they can progress from following teacher blindly, to selecting their own parameters, class intervals, scales. Particularly in human biology, such work has high significance and relevance and gives a stimulus to some complex mathematical skills of which many pupils are capable, if motivated.

Worksheets result in lack of practice in visual communication. When lack of muscular co-ordination is added to deficiency in natural ability, not many adolescents can draw well; that those with natural ability should feel the glow of success the whole time challenges the teacher to train the artistically less able in the elements of recording observation by diagrams in a series of easy stages. (See references 17, 22, 52 for further ideas.)

At the simplest level, pupils can identify parts, and write their function on a prepared diagram. Next, a part of the diagram can be left undone so that they can fill in one or more simple parts themselves from observation. Drawing can be done co-operatively – all attempt to observe and draw, or each group may be given a particular task. The best attempts are selected by the teacher and put on the board by the pupils themselves (although this may lead to embarrassed giggling it is often a useful safety valve after hard work!) or, for the sake of speed, by the teacher, and the rest of the class asked to suggest or make improvements. Those who excel in observation rather than draughtsmanship thereby experience success and the whole class has some reason to contribute and learn. Finally, the amended drawing is put on a master sheet by a good draughtsman and duplicated for the benefit of all. The names of the major contributors should appear on the finished product. (Felt pens are expensive but give added facility and motivation, for those who find drawing difficult, to add to a wall display or exhibition.)

Worksheets are impossible in non-streamed groups. If worksheets are felt to be the solution to a particular piece of work, groups may be structured so that the able help the less able in comprehension; progressive stage worksheets may be used – the simplest beginnings will be comprehended and covered by all, furthermore difficult work for faster-paced pupils comes later and may extend into project work.

Various reasons given for not using worksheets
1 *It takes a long time.* So does the construction of any learning situation; teaching off the top of the head is pleasant for a change, but when used 100 per cent of the time hardly constitutes the behaviour of a trained member of a profession!
   Worksheets and workcards free the teacher from wasting his energies on educationally unproductive repetition of basic ideas and allows time for the essential individual tuition.
2 *The laboratory technician can't type,* you can't type and your writing is unreadable. You are excused until you learn to type (fast!).
3 *You've run out of paper.*
4 *Pupils love/hate them.*
5 *Pupils lose them* – use workcards.
   Finally – no one method ever succeeds all the time – although I must admit I've seen classes where any group discussion leads to riot, working meekly and well under instructions from a worksheet, while ignoring the instructions of the teacher!
   Teachers thinking about the general use of an independent learning approach with older pupils should see Carré.[21]
   The above description should be used to help in the critical and differentially paced use of this approach at all ability levels.

A variety of methods

Examples of different types of worksheets

Record only
'To which stimuli are you sensitive?' (Figure 21)   This sheet was used with the Nuffield Secondary Science Theme 3 work on senses, where the class are asked to decide the difference between two subjects which appear similar at first sight, say what difference in stimulus is presented to the brain, and with which sense organ it is noticed. The class contained non-readers and the verbal ability and vocabulary level was low, so help is given in describing the objects *and* the stimuli. The class saw the pairs first, then the sheet was given out after they had attempted to verbalize the differences – thus they had some kind of concept in their minds on to which the verbal label could be stuck. Most of the answers were given as a check to their own correct observation; A was completed as an example, I and J were not experienced in the laboratory and so the pupils were asked to perform the simplest possible level of experimental design.

It is incredible how many differences such pupils will find – and the work has the flavour of a party game without some kind of structured discussion; the sheet helps to provide this on the basic question 'Have you only got five senses?'.

Fast finishers were asked the question 'What are you *not* sensitive to?' and given clues.

To widen perception
'How fishes swim' (Figure 22)   This was used with an able group of 16-year-olds used to constructing their own record; there is, in fact, no reason why this should not be a workcard.

Most of the pupils had seen goldfish, they felt they knew what they looked like, but this detailed series of questions, too many to write on the writing-board, was perhaps the first time the pupils had looked in such detail at this familiar object.

It is also a good example of a lesson which requires much thought and little apparatus.

Short steps
'The Bunsen burner' (Figure 23)   Another example from *Quest,* exploring the working of a Bunsen burner in steps suitable for less able pupils. In the original this was typed using an especially large 'Jumbo' type to aid slow readers.

Independent learning
'Transpiration' (Figure 24)   The excessive use of paper is valid here as the class was a mixed ability fifth form, needing much individual tuition. Such programmes take a good deal of time to prepare, but in this particular school a bank of such sheets has been built up over the years.

Which stimuli are you sensitive to?

FILL IN THE GAPS   SOME OF THE OBJECTS ARE HERE TO HELP YOU

OBJECT	ONE	TWO	STIMULUS	PART OF BODY
A cards	White	Whiter	Brightness	Eye
B cards	Grey	------	Brightness	------
C cards	Red	Green	-------	Eye
D cards	Circles	Triangles	-------	Eye
E bottles and cotton wool	-----	-----	Movement	Eye
F matchboxes	-----	-----	Loudness	Ear
G whistles	High note	Low note	Pitch	------
H glass squares	Smooth	Bobbly	-------	Skin
I -----	Hot	Cold	Temperature	Skin
J -----	Sweet	Sour	Taste	Tongue
K bottles	Nice smell	Horrible pong	-----	Nose
L matchboxes	Heavy	Light	Weight	------
M envelopes	-----	-----	Touch	Fingers

> N.B. A — given as an example
>
> I, J — as much to ensure comprehension of column 1, as for VERY simple experimental design.

HAVE YOU ONLY GOT FIVE SENSES???

FAST FINISHERS

What are you NOT sensitive to?

> N.B. 4 stations provided showing:
> 1. Magnetism
> 2. High pitched sound — dog whistle
> 3. Bones etc. hidden by things above them
> 4. Things happening at high speed

What do you use to detect it?

> Answers expected — but given only in discussion:
> Compass
> Oscilloscope
> X-rays
> High speed film and flash

HOW MANY MORE CAN YOU ADD ON HERE?

Figure 21

A variety of methods   103

How do fishes move?

This is an exercise in OBSERVATION. A good scientist can record his observations clearly and accurately. Each group is provided with a goldfish in a vessel of water. Follow the instructions carefully answering the questions by referring to the movements of the goldfish. Draw simple diagrams where necessary.

Remember to use the facts we established last week when we dissected the muscles of a herring.

1. What propels a fish through the water?
   Fins, yes! but which fins do what?! How many fins does it have? Are they all the same shape? Are they all in pairs? Are they flexible? or supported? or both?
   Draw a simple diagram of the goldfish to show the position of the fins. Describe why the movement of fins should propel the goldfish forward. Is the shape of the fish important? Why?

2. How does the goldfish move forward?
   Investigate which fin or fins are responsible for propelling the fish forward. Which fin is mainly responsible for this? Does the body move in a similar way to the fin? (Look from above - remember the results of your dissection - make a hypothesis.)

3. How does the goldfish rise and fall in the water?
   Which fins are responsible for a vertical movement in the water? (N.B. Several of the goldfish are in suitable tanks to help observe this.) Can you see how the fins control this? Is the tailfin involved?
   Describe carefully how you think the fish achieves such a movement. Observe a fish that remains motionless in the water. How does it achieve this? Which fins are responsible?

4. How does a fish maintain its balance?
   Fishes are 'suspended' in water. If the water is rough or turbulent, how does the fish maintain its balance? Which fins compensate for 'roll'? Which fins compensate for 'pitching'? How does a fish change direction?
   Carefully explain using diagrams.

5. How do the following fish move?
   Flatfish, seahorses, eels, flying fish (how do they fly?), mudskippers.

6. Why don't fish float - or sink for that matter?
   Find out, next week we'll talk about it.

    Rolling   Yawing   Pitching

Figure 22

| 004A | QUEST | 00401 |

The Bunsen burner – part 1

Study the 'Bunsen burner' card.
Fix live match in burner mouth like this:

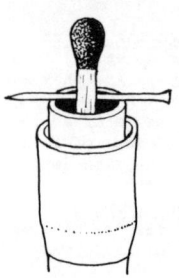

Watch very closely.
Open regulator and light the burner.

Does the match light at once?
You might have to try again with a new match.

Stick a 'Bunsen burner' picture in your book.
Label the parts.

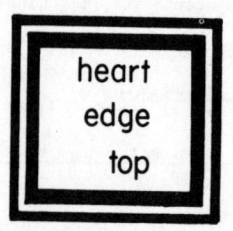

heart
edge
top

I fixed a live match in the _____ of a roaring bunsen flame. The match did not light at once. This shows the _____ of the flame is not so hot as the _____ or the _____ .

Figure 23

A variety of methods

Transpiration

Last time you saw that water evaporates from leaves and this loss helps the roots to take up water. This loss of water from the plant by evaporation is called TRANSPIRATION.

LEAF SECTION   Label

What is this structure?

....
....

Where is it found? ....

....

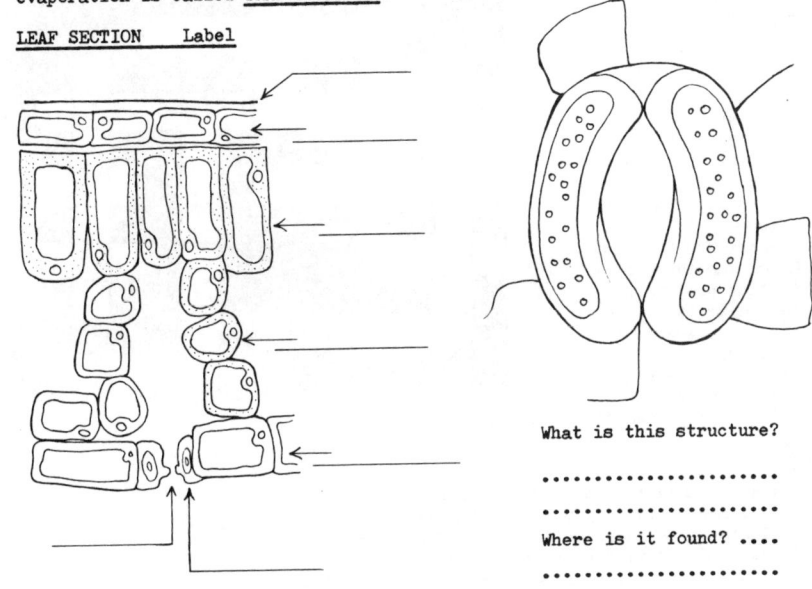

* ALL OF YOU DO THE EXPERIMENTS MARKED WITH AN ASTERISK

The rate of transpiration can be measured using a POTOMETER (one has been set up). As the leaves transpire, what would you expect to happen to the water level in the potometer?
Readings are being taken at regular intervals so that the rate of transpiration/given unit of time can be measured.

1. Another type of potometer is the weight potometer. This is to measure the loss of weight due to water loss.

You will need:- a balance, potted plant, polythene bag, thread.
Cover the pot and soil with the polythene bag, tying it together around the stem of the plant. Place pot on the balance pan and take the initial reading. At the end of the lesson take another reading.
Leave until next lesson.
What has happened to the weight?

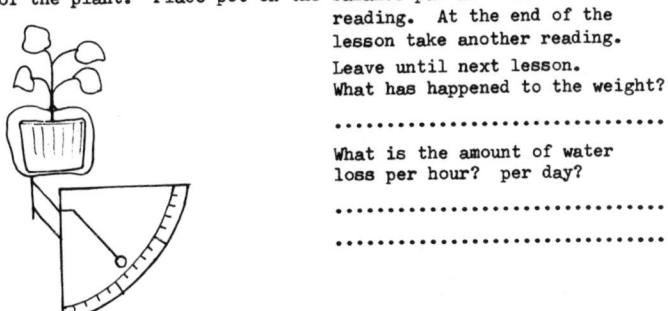

....

What is the amount of water loss per hour? per day?

....
....

Figure 24

Crossword - example by a 14-year-old girl

Across

1. Diabetics inject insulin with this.
2. 1 tonne of beef pancreas make enough of this for 2,000 diabetics to live on for a year.
3. The main source of energy within the human body.
4. The tablet test for ketones.
5. Diabetic or insulin?
6. Green vegetable containing protein and carbohydrate.
7. & 9 Discoverers of insulin.
8. Diabetics rinse and store syringes in this to keep them free from bacteria.

Down

1. Diabetics need to .... their disease.
10. If you are overweight you have a greater risk of getting this disease. In India it is known as the 'Rich Man's Disease'.
11. Produces digestive juices and insulin.
12. When the body chemistry of sugar goes wrong, fat produces these.
13. When diabetics have this, they need some sugar quickly.
14. Artificial sweeteners almost ..... like sugar.
15. Diabetics must have regular X-rays because they are easy victims to this disease.

Answers

Across 1 Syringe 2 Insulin 3 Carbohydrates 4 Acetest 5 Coma 6 Peas
        7 Best 8 Spirit 9 Banting.

Down 1 Study 10 Diabetes 11 Pancreas 12 Ketones 13 Reaction 14 Taste
      15 TB.

Figure 25

A variety of methods

Games and the application of knowledge

This is no place to go into games theory in detail and depth: students are recommended to read the works of Rex Walford.[104, 111]

By simple imitation of games which are well known, useful learning situations which educate as well as entertain can be devised. The trouble with many biological concepts and facts is not in the initial learning but in their application – situations occurring for this purpose in everyday life are relatively rare and provision of them by simulation and gaming is a good consolidation manoeuvre.

Crossword puzzles
(See Figure 25.) Useful for establishing new vocabulary and technical terms – teachers tend to favour a Travel Scrabble board for making them up, although there are some logical geniuses who can do without such aids. Classes equipped with Travel Scrabble can make up their own – an even more highly motivated form of consolidation, in particular for spelling, so beware and have a biological dictionary handy.

A game based on 'Snap!' (Figure 26)
Lower school work on classifying animals into phyla can be consolidated by classes making their own sets of twenty-two cards – five fish, five amphibians, five reptiles, five mammals and two question mark cards – although any number could be used. The cards are shuffled and each player is dealt five –

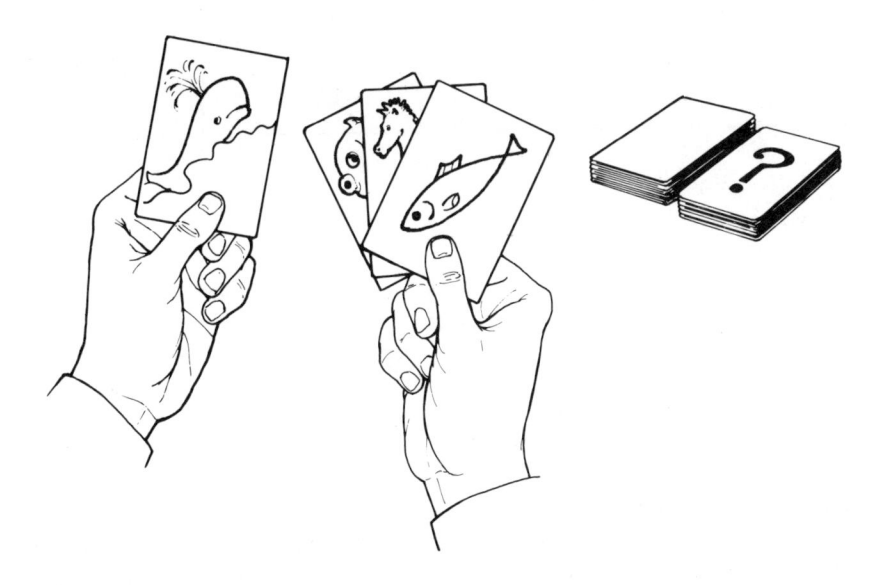

Figure 26

the rest of the pack is placed face down with the top card turned up. If it is a fish, the first player with a fish in his hand covers the discard pile with it. If no player has a fish, any player with a question mark card can then lay it down and name the group he wishes to be played – the object being to get rid of the cards in his own hand first. If a player cannot follow suit or has no question mark card, then the top card of the pack is turned up. If it is the same suit as that on the discard pile, it can be laid on it; if not, it must be taken into the player's hand and so decrease his chance of getting rid of all his cards.

A game based on 'Happy Families (Figure 27)
One particularly compulsive lunch hour waster was a 'Happy Families' game aimed at learning the floral families required in an examination syllabus. This is almost the only possible technique for such work (apart from a great deal of field work), and the comments and competition of the game situation not only enhance learning but help the students realize that they are not alone in lack of knowledge of the terminology. The concept of Scrophulariaceousness is gained more rapidly from this technique than in forming it, even from beautiful visuals in, say, Keble Martin's flora.[68]

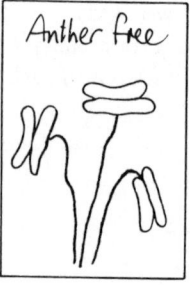

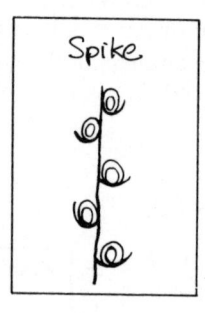

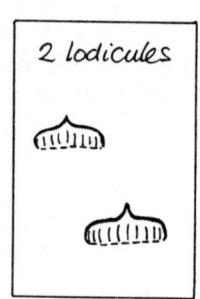

Which family?

Figure 27

In this game, sixth formers could make their own cards depending on which families were set for study and, since many characteristics are shared by two or more families, construct a set of seven characteristics for each family, which together characterize members of that family only. It is essential, as this is a learning situation *not* a consolidation, that a cheat sheet should be provided! (See Table 5.)

Commercially available games
There are many American games[99, 104] available, some of which are useful if complex, needing a high degree of verbal ability as well as logical thought.

A variety of methods

Table 5 The Flora Game: Cheat Sheet

Family	Leaves	Stipules	Inflorescence	Flower	Calyx	Corolla	Androecium	Gynoecium	Carpels	Fruit
Ranunculaceae	Alternate net veined palmately divided	–	Racemes or pannicles	Acti	Free 4, 5 or more	Petals free 0–15	Spirally arranged, free	Superior	Many free parietal placentation	One seed – achenes Many seeds – pod
Rosaceae	Alternate palmately divided net veined	+	Variable	Acti	Free 4 or 5	Petals free 4 or 5	Many, whorls, free	Superior or inferior	1 or many, free or united	Achenes fleshy fruit
Compositae	Alternate or opposite simple, net veined	–	Head or capitulum	Reg. or irreg. disc ray	Pappus scales bristles	5, united tubular 5 lobes	5 joined	Inferior	1 basal placentation	One seed, achene hairy pappus
Scrophulariaceae	Alternate undivided net veined	–	Racemes or pannicles irregular	Zygo	5, 4 united	Petals 5, 4 united	2 or 4 in two pairs or 5	Superior	2, united axile placentation	Many capsules 2-valved
Liliaceae	Long, narrow parallel veins	–	Racemose or solitary	Acti	(Perianth petaloid 6 segments free/united)		2 x 3 free or joined	Superior	3, united axile placentation	Berry 3-chambered capsule
Gramineae	Linear parallel veins	–	Spike, pannicle ears, spikelets	Mono	(Perianth reduced to 2 scales)		3 anther free	Superior one ovule 2 stigmas		Grain

The cards produced by the British Diabetic Society[126] to help diabetics learn the ins and outs of a low carbohydrate diet can be used in many types of game, a very flexible and usable production.

'The Great Blood Race'[100] is a good consolidation of work. It is applicable to mixed ability classes as long as certain ground rules are observed.

(a) One member of the group needs to be the sort of pupil who can read games rules. As this is not one of my abilities, I was glad of the help of a 13-year-old boy when playing this game.

(b) The terminology used needs at least one good reader in the group, and teacher's help on pronunciation will also be essential. The game provides excellent consolidation on the use and understanding of new words.

(c) The game as it stands is too complex and it is suggested that it be taken in stages. At first the pupils play straight round the board – which shows the whole path of the body circulation – before taking Chance cards which can land the whole group in the blood bank with no hope of getting out! Also, pupils need to play the routes through the stomach, kidneys and liver and hit the arterial venous anastomosis, which they may not do if the game is played as the rules suggest. Having got the idea of circulation, then the game can be played in full, bringing in all the hazards which the blood needs to sidestep if a steady state is to be maintained.

Perhaps the best recommendation of this game is that, although the 13-year-old boy had 'done' the blood he learned a lot, and a girl with a short-term-memory handicap learned about blood-clotting mechanisms with remarkable rapidity and retention. The rest of the group, like myself, just enjoyed it!

Pupils as teachers

You never really learn anything until you have to teach it. This piece of folk truth is commonly heard. One girl from a famous university with a broad-based and excellent degree in biology, said she had never learned so much as she did on her postgraduate education course – about biology that is, not about education. She had, in fact, learned little new, except where she was unsure of the logical sequence of the facts; and also, perhaps, she learned to ask questions. What she certainly had learned was how to organize facts, and (here we need to use a bit of education jargon, with apologies to those who are allergic to it) arrange them into 'new and meaningful wholes'. She had also to verbalize much more than she had, even in a university famous for its tutorial system, and this also had had a learning effect. Pupils can benefit from both these experiences.

When the efficiency of learning by teaching is thus demonstrated every year by the new intake of student teachers, why is it that they rarely apply such principles in their attempts to create learning situations for their pupils? Some of the more obvious uses of *pupils as teachers* have already appeared in the

A variety of methods

preceding sections – pupils enjoy giving demonstrations, pupils often get more impact out of marks given by the judgement of their peers than by teacher's mark, pupils can educate each other discussing the answers to worksheets and clarify ideas, techniques or mathematics for each other. Where schools set their own syllabus and examination for external moderation, further strategies are possible. In one school pupils even chose their own syllabus from an outline giving details of most of the eight themes of the Nuffield Secondary Science project. In another school, the pupils became so involved in aerodynamics that a special paper had to be set, as the whole of one term was taken up with this to the exclusion of all other aspects – biology–physics is more interesting than physics–physics? Luckily the particular examining board was a very well-organized institution and the final version of the paper had to be in only a fortnight before the examination.

R. H. Tawney[103] once said, 'I can never be sufficiently grateful for the lessons learned from the adult students whom I was supposed to teach, but who, in fact, taught me!' Here is one student teacher's report on his own experience of pupils as teachers in a sixth form.

The first possible method of involving pupils as teachers can be described as follows. Scene one involves a double period of consolidatory work on sixth form 'Digestion and Absorption'. The nine strong group had previously carried out three periods of self organised practicals, and completed set homework which served as the only introduction to the topic. The class had been divided into two groups and asked to analyse the results from the practicals and provide spokesmen to relate to the other group the procedures, techniques and problems encountered. This then served as a basis for discussion on the subject involved, with each group having to exchange their conclusions and provide reasons for any statements made. This approach provided a full-blooded debate, with individuals even arguing with their spokesman, and members of the opposite groups provoking further debate. The spokesmen were then allowed to provide conclusions.

Thus for five whole periods the teacher involved had acted as tutor, laboratory technician, chairman and mediator, offering advice when needed, expanding areas of thought, but not actually teaching in the accepted sense of the word. This approach proved to be extremely useful provoking pupil orientated problem-solving, applying the various principles involved as well as providing consolidation. The first task to complete was an outline summary of the main aims, and provision for further reading.

Scene two again involved the subjects of 'Digestion and Absorption', but more specifically, dealt with chromatogram analysis. The pupils were again allowed to organise themselves and proved capable of doing so. In this case the 'pupil as teacher' method involved asking one of the group to act as 'expert' in the field of chromatography. The pupil chosen had previously been involved in a project along similar lines to the work in question, and hence all enquiries and problems were therefore referred to him, rather than the teacher. This again proved to be a dynamic approach, with the pupils providing their own 'drive and initiative' in their own specialist areas.

It may always be worth bearing in mind the interests, hobbies and knowledge of individual pupils and being willing to draw upon it for the benefit of the class and oneself.

Perhaps one other experience is worth relating at this stage, concerning a discussion period with the now familiar subject of 'Digestion and Absorption' – but more directly concerned with 'Diet and Dental Decay'. This topic had been set as homework and was based on text book data and questions. This time I had purposely refrained from researching the topic, and hoped that the discussion would be a group effort including myself as one of its members. The result was not very encouraging but such an approach could, I'm sure, be made to work in other circumstances.

These techniques have all been related to older pupils but they could and should be possible to adapt for lower age groups, and be successful if handled with appropriate care. It obviously depends on individual classes and on the teachers concerned. Planning and preparation would be essential.

Methods of teaching older pupils (16 to 18-plus years)

It is reasonable for teachers of older pupils to aim for the best possible public examination results, but it is short sighted and damaging to the pupils' best interests if the teacher insists on making life easy for them by teaching methods which require little of the victim but sponge-like absorption power.

Most examinations in biology these days require powers at least of critical thinking, particularly with respect to experimental design, and analysis of possible conclusions to be drawn from evidence. Should the pupil finally reach the desired nirvana of higher education, the lack of skills essential to make full use of what is offered there leads certainly to anxiety, if not to frustration and failure.

No one method is advocated, nor is any one method forbidden; what is essential is that the methods used help to give the pupils the necessary skills, and confidence in their own abilities and aptitudes, to overcome their disabilities. Better that a pupil should discover, at this stage, that higher education is not for him, than to make the painful discovery when he is actually in the middle of it.

Taking concise notes

There seems to be little point in dictating notes to pupils at this stage. If they really cannot put things in their own way at first, a written sheet should be given with comprehension questions, either written or verbal, followed by a series of paragraph headings by which they can make their own notes. Dictated notes are taken automatically, and pupils get into the bad habit of mental inertia, instead of systematically thinking about what is being said and writing down a simple sentence describing what the message means to them. One of the commonest criticisms of university first-year undergraduates is that they have *no* idea of how to take notes and find it impossible to take down every word coming from the lecturer.

Taking notes critically also allows the pupil to be aware of things he doesn't understand; these are all too easily recorded as nonsense in dictated notes. It encourages skills of making a précis and of linking what is being said with what has been said previously, and is altogether a different and more efficient learning process, not only for those proceeding to higher education but for all other members of the class; and, of course, the process should be started with younger pupils.

Conscious teaching of these skills should allow the teacher eventually to proceed to a university teaching mode in the last year of secondary school.

Seminars
One method which can be used in schools involves the written sheet with comprehension questions referred to above; several of these sheets can be used on different aspects of a topic, and one or more given to each member of the class to prepare and to explain to the rest in a seminar-like session. Later, pupils can prepare their own seminar topic. What is important here is not only the exercise in comprehension but also the use of a principle much used in higher education: that *pupils can educate pupils.* Teachers seem to love exhausting themselves rather than letting a single pupil get his foot in the I-can-explain-it-better-than-you-can door.

What about study guides? Whether you are using Nuffield A-level Biological Science[77] or not, there is useful material in their Study Guide for consolidation and application of basic principles. But it is tedious and mind-boggling to go through a study guide asking each one to attempt to untangle each question – far more to the point to vary the activity by setting some for preparation by each pupil who then presents them to the group. Thus each individual in turn can replace the stress of having to answer with the delight of criticizing someone else's answer.

Independent learning
This is where some universities are ahead of schools in biological education.[53,107] Systematic planning of knowledge, followed by a careful analysis to see which is the best method of transferring that knowledge to the recipient, has resulted in a variety of programmes using a variety of methods. For example – when one studies a locust should one use preserved specimens, live ones, film loops, a set of slides with accompanying tape, histological slides, 16 mm film, book references, historical aspects? What does finance permit? What does time permit? What will libraries permit? The answer is often to develop an independent learning approach, selecting a variety of materials and methods. Some schedules are worked through individually, catering for a wide variety of individual learning speeds, others are designed for group work, where, for example, there is no point in each member of a group slogging away through each investigation, but all can benefit from seeing each other's work. This is an application of the pupils-educate-pupils principle to practical work; it also saves time, temper and laboratory bills! Obviously some parts of the material might involve the showing of a film, for example, to the whole group; there might be a group seminar, but there is also the necessary individual tuition both as a final consolidation and also as a follow up during the ongoing study

of the schedule. This again provides a learning situation for the social and verbal skills needed during an individual tutorial at university; these are often learned by the third year in college but would have been even more valuable had they been usable in the first year. What the student needs to learn is firstly how to diagnose the help he needs and then not to be too shy to get it.

Libraries and booklists
If social skills are well developed, then a student at university could actually ask his lecturer which of the twenty-seven references given (three in German, four out of print, ten in journals at the binders and the rest available in one book the only copy of which was stolen last year) are really essential, or which chapter or even sentence is the piece of grain at the bottom of a bushel of chaff. One suspects that many booklists have been given, both by students and staff, merely as a façade calculated to impress the reader with inflated ideas of the amount of reading done by the writer.

Facetiousness apart, I have had to cope with students who had been sent by their biology teachers to the school library and to local libraries to hunt for books which they can never find; excellent training for university, I suppose, but at least the students should know how to tell their mentors, ever so politely, that the task set was a waste of time. The inclusion of passages from books and papers into the independent learning approach is advisable, while help and advice from the librarian is always welcome.

Other resources
Why worry chasing non-existent and irrelevant books when TV is in your living room? Older pupils should be encouraged to produce a file of records of biological programmes seen; the previous night's viewing can form the material for a seminar. This is an excellent exercise in precise note taking, and formalizes what has become an informal exercise. It also helps the biology teacher to remember to see the thing next time it's on!

Class film viewing should also be recorded – otherwise many may switch off and have a quiet sleep; active involvement in film viewing is as important for learning efficiency as active involvement in group discussion.

Written examinations
Multiple-choice tests, comprehension papers, short-answer questions may come and go, but the essay type goes on forever. Biologists still emerge as graduates from universities without the ability to organize an essay by putting ordered thoughts to paper. Older pupils emerge from school examinations cursing themselves for knowing things which they did not write, for choosing the wrong question and for getting off track in the right ones. Schools need to give practice in how to deal with an essay question paper; the following lines are suggested as basic, and for practice in school:

(a) *Recall check*   Begin at question 1 and write down briefly what you know about it. Do this for all the others. As you write down information for question 5, something may be recalled about question 1 – the main problem about biology examinations is to *recall as many known facts as possible in the given time*. It is well known that recall occurs often when one is *not* trying to recall; hence the shift of attention to question 5 pays dividends in recall for question 1.

(b) *Selecting questions*   Re-read the instructions at the head of the paper as to how many questions in which sections you must do, then make your selection. *Never select your questions without doing your recall check* – you may waste time writing a mass of stuff about the first half of the question and then find you know nothing about the last half; and it's no use kidding the examiner you ran out of time – that one is as old as Aristophanes.

(c) *Organizing the question*   Now arrange the wild selection of scribbles under each question into order – select a heading for each paragraph you are going to write and then write about it – use it *as* a heading if you like, but it looks more stylish to make it into the opening sentence. Say what you want to say about it, then stop; padding irritates examiners.

(d) *In case of need*   If you don't have time to write up your notes – put them in with your papers nevertheless. The examiner may choose not to mark them, but at least he will see enough to decide which side of the distinction or pass/fail line you should be!

It should be obvious to teachers that such instructions cannot be implemented overnight without previous exercise of the skills involved. It is a good strategy to set in class, and mark, or at least discuss, paragraph headings on essay questions, apart from the fact that it provides a good means of evaluation for the teacher and consolidation for the students. The use of flow diagrams appeals to some pupils as an aid to organizing thought.

It is also a useful ploy to ask the class to mark each other's essays, given a plan of marking for facts, explanation, application of theories, organization and synthesis, complete/incomplete data retrieval.

Practical examinations
It seems incredible that these still exist in biology today, but while they are with us, it is necessary to prevent our pupils from panicking and help them to do their best in an impossible situation. It might be a good idea to provide a means for the group to relax and recover after the ordeal. It is an arguable and tenable hypothesis that A-level biology examination practicals take more out of the teacher than the pupils.

Those practicals which still exist seem to have jumped onto the bandwagon of 'scientific method' attitudes to experimental work, and they do expect critical analysis of experimental design, selection of viable hypothesis and application of basic principles – all much more appropriate to the calm of the written examination than to the hurly burly of the practical. Whatever one says about Nuffield A-level, the structure of the examinations is to be admired.

If candidates are to have any chance, they must use these methods of thinking throughout the years preceding the examination. The shock of meeting them in an examination room for the first time, and in a practical situation, too, is destructive.

The actual practical work itself is not very demanding technically, and usually full allowance is made for biological variation, such as dead *Paramecium,* rotting herrings which cannot produce pH changes throughout their alimentary canals because the whole lot has undergone autolysis, and beans which have inconsiderately germinated way beyond the length required by the examiners.

Drawing
To spend hours and hours drawing during laboratory sessions is obviously of little use in the practical examinations of today, where the pupil is more likely to be asked to interpret an electron microscope picture. Nor is there any point in spending time looking down a microscope and then drawing what the textbook shows. An *interpretative* approach to observation, and drawing used only where words and photographs will not suffice, is essential; there is more point in saying how a slide differs from that shown in the textbook, and why, than in learning only to feel cheated when the slide given is unfairly not like anything previously seen.

Monolithic textbooks
Some are good, but too often they tell the pupil everything he does *not* want to know and stop short at the crucial point. Many of them were written before sixth-form teachers began to take feedback, and are a monument to the silent conforming multitudes who learned in spite of their repetitive-type teachers.

Do not therefore refuse monolithic textbooks but treat them with critical caution. Take the advice of friends; there are even pundits about, whose Nuffield A-level texts are annotated with what is and is not a waste of time!

What is the alternative? Not every teacher can make up his own schedule for every piece of work – many specially produced series for this level are either deadly dull or far too complex. What is really needed is an exchange and mart of suitable programmes, each one produced by a front line teacher and vetted by an expert.

Lecturing
Some would-be teachers still begin education for teaching with the image of a lecturer in their minds, and feel quite disappointed when, even with older pupils, this method is frowned upon. For the complete argument as to why they should explore all other possible methods before deciding on a lecture,

see D. Bligh (1972) *What's the use of lectures?*[14] The best lecturers fall basically into two groups – the straight showbiz merchants who are *agents provocateurs* rather than handers on of dogma, but who too often enchant without educating – 'Yes, it was a great lecture, but I don't really know what he was on about.' Secondly the straight synthesizers who, from a comprehensive survey of a field, give an overview which educates the novice not only about what is known, *but what is unknown.* This is what university lecturing is supposed to provide – experts at the furthest frontier of their specialism who can give confident predictions of future trends or admit total ignorance and create intellectual excitement. Unfortunately many at the frontiers of knowledge cannot communicate it, staffing of departments often means that a lecture course has to be put on by someone pressurized into it who, at worst, only regurgitates textbooks. The feeling still prevails in many universities that because all lecturers have been at school, they therefore need no further study in education, either its techniques or its philosophy.

But that is the gloomy side of the picture – many lecturers now take professional courses which vary from simple ones on how to put it across to the more usual ones on *why* they are teaching the work as well as *how*, including methods of assessment and objective analysis of both the educational aspect and the administrative aspect of the course. New teachers should take their models from those who temper the academic image with critical insight, and keep the following points in mind when considering lecturing:

Is this the best use of student contact time? Is it best spent in giving new facts or in diagnosing difficulties of understanding and/or application?

Students' concentration is high in the first twenty minutes, but then a change of activity is needed – discussion, filling in a check list on the first twenty minutes' work, seeing a film, doing a problem. If a lecture is broken up into three twenty-minute sessions containing different activities, learning usually benefits. However, there are still many university lecturers who are unaware of this truth, so perhaps they should be given training in coping with fifty-minute non-stop lectures?

Lecturing notes and nerves   These often form a vicious circle; under the stress of a large sixth form, new teachers feel they must adopt the impossible role of the all-knowing academic. Notes which seemed perfectly adequate the night before now appear unreadable. Some counter this by making a list of points on which they are going to expand, in large black felt pen; others cling gratefully to the overhead projector, for convention allows them to *look down* at this! One professor felt that unless he could draw complex biological drawings on the writingboard during his lecture he would lose face in front of his students – so even the mighty can get hangups in this situation. On the other hand it must be admitted that an ambidextrous botany lecturer used to provoke loud applause as he drew *both* sides of the diagram simultaneously.

Lack of feedback This is increased by the factors described above – the lecturer who is nervously immersed in his personal fright and lecture notes takes little or no feedback, and drones on to the boredom of all.

Analysis Even when lecturing, knowledge the students are assumed to possess will need to be assessed; matters explored, and the best way of doing so, must be thought about, as must methods of evaluation and consolidation. Add to this the need for logical sequencing and synthesis of an overall view of the topic and it should be apparent that a good lecture takes a great deal of preparation and that a piece of independent work for the class followed by a consolidatory discussion would be perhaps not only more appropriate but actually easier.

The demagogues Few students listen meekly to dogmatic lectures today – and there may be problems with one or more members who take the floor with any or no provocation and bore everyone else with their irrelevant ego trip. Those new to lecturing are at first delighted that someone is actually taking enough notice of what they say to argue with them! But on the whole it is better, after an airing of the grievance, to ask the demagogue to write and present a paper; then he will have to discipline his thoughts and all may benefit from the results, even the lecturer.

Most groups of students are mixed ability groups, and the lecture method is not well designed for this learning situation. Nevertheless, for some personalities and some topics it can still be appropriate when used judiciously.

Finally, a reasonable use of an independent learning approach for older pupils will be found in Appendix E.

5 Difficult or impossible to teach?

Difficult classes – check-list of possible reasons

The children are unteachable

Because of *personality differences* between pupil and teacher; some will never be anything but incompatible. It is important that a pupil should meet all kinds of different personalities in his teachers; they identify with some and reject others.
Because of six years of conditioning that school is boring and irrelevant and teachers are to be looked down upon as moneyless, sexless, smug goody-goodies.
Because the pupil is a *genuine psychopath* with or without a long child guidance record.

The second and third types can disrupt the learning process of a whole class and tire out even the most experienced teacher. Teachers should have the right to remove such pupils from the class, but it is surprising that some schools don't face this problem in a practical way. There the teacher is made to feel inefficient because he cannot cope with kids. There is no reason why he or the rest of the class should be made to suffer for the past mistakes of others, whether it be parents or teachers; what is more, it is a waste of public money and a diminution of the rights of the rest of the class.

The structure and organization of the school is anonymously bureaucratic and insensitive to the needs of the pupils

Too often a school will select as its criteria for success examination results and other of the superficial attributes of success. Hence pupils suffer watered-down academic courses. Other factors may be:
(a) Where the exam-passing, fact-cramming attitude prevails, there may be little insight into the pupils' needs and attitudes. All must fit the mould. Urban society tends to *anonymity,* and where this also prevails in school it is unreal to expect a pupil to achieve any degree of socialization.
   This is not a necessary concomitant of comprehensive education – at least large schools are aware of this problem and attempt to ameliorate it

by tutor groups, year heads, etc. In a school of 400 where the head teacher claims to 'know' all the pupils, one suspects that the knowledge can only be superficial.
(b) *Interpersonal communication* between pupils is discouraged except in breaks where pent-up emotions are expected to be released in the regulation fifteen minutes. There are schools where silence would seem to be an educational objective. Conversation between pupils can be a structured part of the learning situation – interest and investigation when promoted are reinforced by verbalization.
(c) *Lack of pupil involvement,* either long-term in understanding of the rationale (very often there isn't one) in their curriculum or in the short-term day to day running of their courses.

The course should not be a supermarket provided by the teacher; pupils' opinions should be sought and their motivations and interest followed up. Where pupils have helped to choose their own routes through the material offered not only more involvement has resulted but also a degree of *self responsibility* ensues.

The teacher is at fault

This may be a reflection on the lack of professionalism in teacher education as a whole, or on the fact that most teachers used to classes of clever children find that in comprehensive schools catering for all levels of ability they have to re-think almost everything when dealing with the less able and have not had either the time or the help with which to do it.
(a) *The teacher cannot communicate* with the class on anything but a superficial level since the class uses different codes of speech, both in vocabulary and in sentence construction.
(b) *The work is irrelevant to the class* since they appreciate neither the long-term aims of examinations and future career prospects, nor the academic delights to be found in the study, say, of the notorious *Amoeba*.
(c) *The teacher's personality is such* that he requires to be the fount of all knowledge and value judgements; he feels he must teach rather than that the class should learn or even learn to learn, and he gives neither the time nor the motivation to do either.
(d) *The teacher uses passive approaches,* not involving the pupils or demanding anything from them but silence. If such a teacher turns to active approaches the class may then resent the awakening from its age-long slumber.
(e) *The teacher neither prepares the lessons nor evaluates them.* Student teachers are often laughed at for preparing lessons and in their first jobs are often told to forget all that nonsense. Experienced and successful teachers of 'difficult classes' are agreed that planning and preparation

Difficult or impossible to teach?

must be meticulous, and that their standards of discipline drop if it is not. All they need is the time for preparation, the facilities for reproduction of resource material, and an active laboratory technician.

The class rejects the teacher's values

The pupils wish to be part of an adult world which is more exciting and which employs them more gainfully than (it seems to them) does the school. They have a fervent desire to leave school as soon as possible and this is often echoed equally fervently by their teachers. These pupils are often of above average ability, and sometimes very bright indeed. For them the criteria of significance and relevance of the factual material are of paramount importance.

The pupils genuinely lack ability

Variation must be taken into account – pupils may be good at languages but poor in mathematics, or good at technical drawing but unable to spell. Judicious use of the able in one field to tutor the less able in others helps both; communication and verbalization improves and the pupil-tutor gains a feeling of success. It is very difficult for the middle-class teacher to praise sincerely the meagre achievements of a less able pupil which nevertheless may represent a pinnacle of heady success to the pupil involved.

Variation also occurs in the *cause* of low ability – there may be family, social or psychological reasons; research has uncovered the concept of the able misfit who is often to be found in low ability groups, under-achieving largely because of boredom.[83]

Slow pace, little retention, low concentration. Sometimes the methods are at fault here rather than the pupil.

It is of prime importance that the pupil should be allowed to express findings and question verbally at first before there is any attempt to write them down. Some may never achieve literacy, but a store of well internalized experience which can be verbalized has a marketable value.

Dull *repetition* should be avoided in favour of a type of reinforcement and consolidation which aims perhaps at practical application, evaluation of evidence and problem solving. More careful consolidation is required here than at higher levels of ability.

Low saturation point for factual knowledge as well as lack of persistence and curiosity may result in lack of concentration which can also be due to emotional factors – many of these pupils are nail-biters, compulsive ball-point pen snappers and wide open to risks of becoming addicted to smoking and other addictive drugs such as alcohol. They appreciate a switch of activity to one making less demands on them – particularly the use of art forms even if only felt pens and paper. They may tire half-way through a lesson or half-way through a term or half-way through a year. Recognize the symptoms and provide relevant work, but of a cabaret nature, not only at the end of the summer term!

Selection of relevant work must not lead to dilution and a resultant structureless mass of fun experiments.

Block timetabling is appreciated by most pupils. It enables them to work at their own pace, and to be able to concentrate on something they are involved in, rather than to have to switch concentration at regular 35-minute intervals. Also individual learning (rather than class teaching) tends to flourish more and the teacher becomes less anonymous.

In choosing *factual objectives,* teachers must accept that a few solidly experienced concepts, reinforced by practical application and consolidated by problem solving are more likely to be retained than a mass of unrelated facts.

In choosing *behavioural objectives,* teachers must remember that Rome was not built in a day. What you are aiming for is:

1 Self-programming, self-disciplining pupils.
2 Science used as motivation towards verbal fluency from visual beginnings.
3 Verbal fluency in both vocabulary and sentence construction leading to written work and comprehension and as an aid to reading skills, and numeracy.
4 Objective observation including such habits of thought as accuracy, realism, openmindedness, persistence and a little logical thought.
5 Concrete experimental design and hypothesis formation as the next stage from observation. Fiddling can be turned to creativity, but abstract thought and hypothesis formation will rarely be achieved.
6 Good discussion built upon good communications between teacher and class. Teachers who have to build upon the conditioning of the class by authoritarian predecessors will find it difficult to establish the idea that talking in class is *about* work rather than an escape from it.
7 Confidence in the pupil that he is allowed to make mistakes so long as he learns from them. Too many pupils are inhibited from even attempting to answer questions because of an unfounded lack of confidence in their knowledge.

Impossible to teach

In spite of good sense, good will and high standards of professional training there will always be some pupils who are impossible to teach all of the time and others who are impossible to teach some of the time.

In one school, recently reorganized, a teacher asked for help with a particular boy; the teacher thought his teaching methods must be wrong, but when investigating the problem it was found that the boy had records from Child Guidance clinics making up a file centimetres thick. There are many such children and teachers are not trained to deal with what Child Guidance clinics find impossible. The teacher should quickly refuse responsibility and attempt to promote action from those who should be dealing with the matter.

Those who seek attention form perhaps the largest group. Even the cane is welcome! One suspects that many who recommend the cane were themselves children living in indifferent families or societies and they also welcomed this obvious signal that at least their presence had been noted, whether for good or ill.

The bored also compose a large group, and boredom engenders the birth of the attention getter. Schools seek to deal with regularly disruptive pupils in a variety of ways – all of which have to be seen in the context of the particular schools but which are listed here for critical appraisal:

Somewhere to relax   One school has a music room where girls can play pop music and dance to let off steam. Few want to do this all day, and if it becomes a refuge for a whole class the teacher has to think long and carefully about his methods!

Other schools provide paint and modelling facilities which also give a good opportunity for pupils' troubles to be sorted out, rather than in a face to face confrontation with either staff or school counsellor. Lest some think this is pandering to pupils, it is amazing how many inner-city pupils manage to cope at all with school given the other appalling stresses and strains in their lives.

Parental pressure   You can try sending a note to a parent saying you will refuse to teach his child if the brat persists in singing loudly throughout your lessons, but the parent may well be the root cause of the problem. Several authorities are experimenting with camp schools for pupils whose disruptive behaviour is largely due to the parents. On the other hand there are parents who are genuinely amazed when told that their son is obviously doing his homework with one ear on the TV, and who will remedy such situations quickly – always try parents but don't be surprised if they do not respond or are just unhelpful.

Segregation   Some schools segregate disruptives for whole weeks or whole days. Depending on the strategies employed this can be useful or totally negative. In one school it was *de rigueur* to spend at least one week per term in the cooler, and as the treatment is symptomatic rather than diagnostic, the cooler class got so large that it is now reserved for the Intensely Wicked which makes it even more the mark of an élite! On the other hand some schools use segregation as a means to put across the idea that education is not a right but a privilege, and where the school curriculum and methods of teaching back this idea up, pupils can, and do, change attitudes to learning, including the fact that even boring learning can be useful. Such classes need expert organization and a great deal of flexibility and thought.

Some pupils segregate themselves – particularly those of 16-plus who may resent being kept at school when they could be earning money. Work may be boring but at least you can see the significance and relevance of it.

6 Biology's special needs

Using animals in schools

While teachers must be glad that biology has moved away from necrology, many are appalled at the indiscriminate killing of animals and plants, which has implications for moral and ethical education, and also for the critical evaluation of methods in science teaching. For information regarding the use of animals consult the publications of the Educational Use of Living Organisms project.[98]

Some points to consider when using live animals in school must be:
Is this the best method for establishing a particular fact or concept? Be honest, are you keeping that boa constrictor just for kudos or is it really used in a way for which earthworms would be inadequate? What are you using that splendid animal house for, except to impress wandering inspectors and to help children escape from human frustrations to the domination of little furry faces?

Having satisfied yourself on this point, then ask *whether you are using the animal as much as possible.* This is especially relevant when killing an animal. Peter Fry (1968)[44] has pointed out that if frogs to be killed for general dissection are 'first anaesthetised by standing in 1:1,000 ethyl m-amino benzoate (MS 222 Sandoz), killed by pithing to destroy the brain and then placed in fresh water for 20 minutes to remove the anaesthetic, the resulting bodies could then be used for (1) heart action, (2) nerve function by removing the sciatic muscle/nerve preparation dorsally, and (3) body structure.' The skeletons can also be used, jaw and tongue movement examined, and eyes investigated. Obviously the use of a deep freeze encourages and facilitates economic use of animals as well as being aesthetically helpful, as for example in the dissection of the dogfish, which is seen to be a beautiful creature when not greyed with sinus-attacking formalin.

Determine clearly in your own mind where you draw the line. An ardent sprayer of roses and murderer of greenflies protests about the killing of *Gammarus* in pollution experiments – is she a hypocrite? There is a continuum in ethics of animal killing; some may refuse to attach locusts to flight machines, others protest about flaying mice skins for genetics, having become emotionally involved with the mice as parents of families. How much genetics in schools is actually taught by breeding experiments as opposed to writing-board and poppet beads? If it matters about mice, why doesn't it matter about

Drosophila? If you are using a live vertebrate in any form of experimental situation make *sure* you are within the law.[114]

Embryos pose a particular problem. Few of us feel a pang when slipping a one-day chick embryo into Bouin's Fluid. It is when an embryo appears to be sensitive to pain, and when it moves and has being, that sensitivities are alarmed. Having to kill a crippled chick after it has emerged from the egg presents another set of decisions to be taken, but in rural science classes pupils breed, rear, kill and pluck their own Christmas poultry. In our urban society we are quite happy to sanction the killing of animals for our own nutrition; as a devoted steak eater I would not wish it otherwise. At least I can claim that I am aware and approve of most of what goes on in abattoirs, whereas in most cases people prefer not to know how they get their *foie gras* or veal. When animal killing was part of everyday life in agricultural communities it still upset people to see their favourite pig killed (see Alison Uttley's book *A Country Child*);[110] but the principle of 'us or them' nevertheless operates, as it does in the case of killing rats and insects which are in competition with us for food supplies. In urban societies death is only met infrequently, and then often violently, and talk about death is often as tabu as talk about sex used to be. There is some educational point in using the killing of a crippled chick as the beginning of a discussion on the quality of life for human cripples, euthanasia, not striving officiously to keep alive those to whom death would be a blessed release, etc.

When slides or film of human embryos at different stages of development are shown to children many are disturbed by the idea that the babies are dead. Showing such material demands careful preparation and communication in depth with the class – and this may lead to the question 'What should happen to babies which die – should we just bury them or use them (with their mother's permission) to show other children how fearfully and wonderfully they are made?'

'Does one inculcate a basic respect for life, but imply that it may be sacrificed in the pursuit of useful knowledge?' This is the question asked by James Limbird and Evelyn Brogden (1968),[62] and it certainly points up the previous arguments and makes clearer the basic premise that facts are often conceptual tools in teaching and not an end in themselves. I think that one can easily and quickly inculcate a respect for life in people of average intelligence level who are morally educated; I think it is difficult to do so in the emotionally and socially deprived, and with this type of pupil I know of nothing which makes a greater impact than work on chick embryos. In a class of 9-year-old remedial pupils I have seen the same exhibition of wonder and awe at a three-day chick embryo with beating heart as I have seen in postgraduates who had previously only seen serial sections. Would it seem unreasonable to hope that the 9-year-olds stoned fewer cats or left fewer overgrown Christmas puppies to starve, because the implications of the complexity of development had been made clear to them? Similarly when using pictures of human embryos, a new

dimension is given to what is too often an unknown part of life, which comes between giggle-ridden ideas on sexual intercourse and the emergence of sanguinary and blue-faced babies from the vagina. It is indeed the miracle of life, as one pupil called it. Therefore I suggest that, after taking careful thought, life may be sacrificed in the pursuit of useful knowledge and in particular in the changing of attitudes. But teachers must be very sure that they have not missed out a stage in the logical development of the argument.

Individual variations in sensitivity. Some pupils (and some quite elderly and tough gentlemen too) blanch at the sight of blood, while some put on attention-getting displays of hysteria. To differentiate between the two is simple; merely put protesters at the back of the class with some interesting work to do while the rest get on with the work in hand. Soon those who are only interested in creating a disturbance will creep forward, while those genuinely emotionally disturbed will remain where they are. Some pupils display show-off ghoulishness with embryos or dissections, others are dismayed to find that they actually like the feeling of the power of death over an animal. Teachers must be aware of the vast range of feelings between these extremes, be prepared to discuss ethics and emotions as rationally as possible and to seek help if a pathological mental state is suspected; this happens but rarely, although it must always be borne in mind. At least biological work helps to diagnose the ghouls who may be in genuine need of help.

The use of abattoir material and animals already dead. Is there in fact a need for live animals to be killed? Is there not a great deal of biology which can be taught using living human beings and complemented with work on lungs, hearts, trotters, etc. from the slaughterhouse? The proper study of mankind is, after all, man and there is a horrid danger 'of uncritically extrapolating the results of animal behaviour to man'. (Tinbergen, 1968)[106]

The use of animals found dead presents dangers of communicable diseases – ornithosis is so widespread among wild birds that today even pigeons from farmers' pigeon shoots are inadvisable for use in schools. The only safe rule is *never* to use such material in your teaching. All material for dissection should come from known accredited sources.

Questions

1 Is it anthropomorphic to think that birds in cages are miserable, that newts in tanks are miserable, that locusts in cages are miserable?
2 Critically evaluate the use of animals in school laboratories or classrooms which you know personally. Impressing the inspector was the reason given for (*a*) a large flight cage for budgies, (*b*) a sea water environment tank – could you have put them to better use?
3 Breeding to kill – should it be left to professional animal suppliers, or be done surreptitiously in school so that pupils do not become emotionally involved?

Biology's special needs

4 Chick embryos – for any given class, state your objectives in using them, and state whether you would (*a*) go no further than day 3, before the sense organs become joined up with the effector organs, (*b*) use the secondhand evidence of a film loop. If you did rear them what would you do with them when mature?

Dissection[92]

The traditional mode of dissection, as verification of what the diagram in the book shows, is almost moribund, if not extinct, and dissection as an investigation using a variety of techniques is much more common.

Pupils often want the status of 'doing a dissection' as they have heard the lurid stories of what goes on in biology laboratories from hyperbolic friends; other pupils hope that it will not happen at all. For the ghouls and the blood haters, therefore, an introduction to dissection as an investigatory technique via a pig's trotter solves a good many problems.
(a) There is no blood.
(b) There are no dissection guides to cast a verification gloom, no puzzling tramline diagrams to comprehend.
(c) Fresh pigs' trotters give life to dull lessons about bones, joints, tendons and ligaments – learning by experience what these structures do instead of the usual verbal and diagrammatic treatment.
(d) Pupils who have an aptitude for dissection can show it; the dissection of the last and most minute tendinous insertion on a toe bone is often produced by the most unexpected pupil.
(e) Even in these inflationary days, pigs' trotters cost little – and many markets will give you them free. Courtesy dictates, nevertheless, that you offer beer money, which will lubricate your supply lines and keep them going.

The following suggestions are the bare bones of the kind of workcards you will need. All this work can also be done using beef bones and joints, but then you will need very large sharp knives to make any impression on the bovine ligamentous joint capsule; one is nevertheless rewarded by large quantities of synovial fluid and acres of shining cartilage.

Muscle, tendons

(a) Scraps of muscle remain – pull on them and note the movement produced.
Q How does one set of muscles produce an entirely different movement from the other?
(b) Cut away the skin on at least the underside of the foot, or, if time allows, skin the whole foot.

(c) With small scissors free one muscle set from its connective tissue and follow down the course of its tendon, freeing that all the way to the bone.
Q How is the muscle joined to the tendon? What is gristly meat?
Q How many branches has the tendon?
Q How is the tendon joined to the bones?
(d) (Optional.) Examine the origin and insertion of an antagonistic muscle – dissecting out the tendons so that you can see clearly how they work against those from your first muscle.

Joints

(e) Bend one of the joints of your trotter.
Using a sharp scalpel, cut across the joint until it opens, then extend the cut around to the other side.
Q Cutting the joint open is hard work – what kind of material is keeping the joint together?
Q When the joint is open, feel with your finger the lubricating fluid. What else helps the joint to move smoothly?
(f) Ligaments (and the ligamentous joint capsule) join bones to bones. Investigate the joint's articulating surfaces.
Q Inside the joint, does bone meet bone?

The questions they ask

Footballer's knee is sure to come up – there doesn't seem to be any housemaid's knee any more! Footballer's knee usually implies a crushed piece of cartilage in the joint which needs to be removed, as body mechanisms for removing large areas of damaged tissue from the knee joint are not particularly efficient. Once the damage has been carved away in a healthy young man, the constantly growing cartilage of the articulating surface will soon provide adequate cushioning.

In the osteoarthritic knee, however, 'like my granny has' the cartilage is not being adequately replaced, and eventually bone may grind nastily on bone.

In the case of gout, crystals of uric acid from a malfunctioning metabolism are actually deposited in the joint between the articulating surfaces – you can model this by putting a pinch of sand grains in the joint and watch them grind the smooth cartilage to bits.

Rheumatism is a more general term and can mean almost anything from ligaments which have turned from the sol to the gel state by lowering of temperature, and thus do not move too well, to a variety of malfunctions of tendons, lubricating fluid, etc.

Sprains, pulled ligaments and tendons can be demonstrated on the trotter, too, and elementary first aid given!

Safety in biology laboratories

All student teachers should during their training become members of a teaching union – they will thereby be insured against the small chance of being sued for negligence in the event of an accident in the laboratory. All students in the UK should be aware of the Health and Safety at Work etc. Act (1974).[55]

In addition to the usual hazards encountered by those teaching the physical sciences,[31, 36] biologists have to be aware of others, such as potential dangers from animals[69] and microbiological hazards.[32] There are other problems such as the ethics of using animals and hysteria about dissection. You must realize when dealing with evolution that such an idea is not acceptable to some adults; and last, but by no means least, there may be parental disapproval of sex education. This last results in publicity, very often coupled with the accusation of 'not being nice', although rarely will legal action be taken.

Most science teachers attempt to inculcate good laboratory manners in their first-year pupils, but judging on a purely personal basis, it would seem sensible that these should be reinforced later, and that teachers should simply refuse to teach laboratory work to those who seem intent only on racing around the laboratory holding lighted splints, or committing sins against expensive apparatus. At the other end of the scale, one teacher doing work on fuels, (where action to be taken in case of fire was taught by demonstration and pupil experience) reported that a fire during practical work caused no excitement – it was merely dealt with by the pupil concerned, while the rest looked on critically at the performance of a well-drilled routine. It seems therefore that routines must be established by practice, and a move away from didactic exhortation in the field of safety seems to pay dividends. Most pupils need to have their perception of possible consequences of their actions extended – the work of the Schools Council Moral Education project on 'Consequences'[70] shows what form such work might take.

Too much reliance on a teacher's presence is in itself a danger – laboratories are exciting places and the absence of a teacher can let creative imagination loose. Do not be surprised if you find a pupil has put lumps of metallic sodium in all the sinks and turned the taps on full. The resultant fireworks are splendid! Responsibility for one's own actions should grow during secondary education through conscious organization of experience diverted to this end, and the work on laboratory safety is a specific case where this can happen. Industrial accidents provide a difficult field for education, probably because the prevalent attitude is that they happen to other people and that safety precautions are for sissies – early and well planned education in school laboratory work might provide an antidote to this in later life.

Familiarity breeds contempt in experienced science teachers too – when student teachers on teaching practice are asked to find out where the main gas, electricity and water supplies to the laboratory can be turned off, many teachers have to admit they don't know, or that the key for turning off the water has long since been lost!

Microbiological precautions are probably the best way to begin to establish attitudes that safety precautions are not for sissies. Most pupils probably already agree that food poisoning is to be avoided; in fact one class who had used disclosing tablets to stain the bacteria on their teeth, and had chewed and swallowed them against the teacher's (unheard) advice, accused her of attempting to poison them as the experiment had coincided with a rampant *Salmonella* in the school canteen. Pupils need not be actually doing microbiological techniques, dissection will do. Perception can be enhanced by using a film loop *Laboratory precautions*[130] which shows how bacteria can be transferred from pencils to mouth, from fingers to toffees, etc. More pupils bisect a heart, then write their worksheets without washing their hands, than have their long hair set on fire; luckily the standard of the Meat Inspectors at public abattoirs is high, and little harm comes, but the potential is always there. Teacher- and pupil-made visual aids on extending perception of possible consequences make a good learning situation, possibly more relevant and significant to the class than those made commercially. Visual aids can also be made on an agreed routine procedure in cases of the various sorts of burns, for example, or the treatment of eyes. The British Safety Council has a useful and authoritative pamphlet, easy to read, on 'First Aid and Emergency Action'.[19]

Finally, new teachers are reminded that they need to admit ignorance in this field – if you have never handled a gas cylinder or performed the hairier heights of chemistry, say so and don't teach it. A nervous teacher is inherently a dangerous one.

Field trips

For safety on field trips new teachers should consult references 4, 30, 96, *and* experienced leaders of outdoor expeditions; although in the case of biology field trips there is more danger to the plants and animals than to the pupils.

The uses and abuses of field work

Abuses
1 Ill-organized denudation of the countryside/seashore. The harassed teacher shovels the class into modern soft-suspension poorly ventilated buses, ill equipped for the delicate semi-circular canals of the pupils. This combines with the effect of too much pop, crisps, sweets – and not enough breakfast (take plastic bags, bicarb solution, which kills the smell of vomit, and moppers-up for child and bus). Spalanzani vomited before breakfast,

Biology's special needs   131

"It's nature trail worksheet No 3 today, Arthur"

but this was a scientific investigation of digestive juices and has little place on field trips. The pupils then spend the day sunbathing or protesting about the rain and bring back loads of sagging plants and exhausted animals, to ignore them and have them thrown out by the cleaners' strike later on in the week.

2 Too many lists and transects. Really, folks, there *are* other things which can be usefully done in field work. Why these are so often done is that the day's field work is too often seen in isolation, and with a total lack of any systems analysis:

   Why do you need to do field work?
   What concepts are you seeking to establish?
   What techniques do you need the class to learn?
   What motivation are you using?

Go back to the principles of lesson planning, analyse the situation and design a series of integrated lessons – allowing for sudden monsoons on the field day itself. How much can actually be done in the field? You can sit all day in a marsh watching the habits of the Lesser Spotted Mugwump, but a film cameraman has probably got it all by two years' hard work!

My own private opinion is that much field work lacks motivation and is often very teacher-orientated, the class merely going through the motions for the sake of peace and acquiring merit. Secondary school pupils are more interested in fashionable recycling studies. Yes, there are some brilliant teachers of ecology – read Nuffield Secondary Science Theme 1 *Interdependence of Living Things*[78] to pick up some ideas (and also see references 12, 18, 25, 26, 29, 49, 58, 67, 73, 113).

3 Ill-taught taxonomy is off-putting. There is no shame in admitting that you cannot use a flora even after a six-week University course on Clapham, Tutin and Warburg – what hope is there for the Sixth? Design a 'Happy Families' game (see page 108) to familiarize your sixth form with some common plant families, or give up and use Keble Martin.[68] Many pupils learn the concept of *Rosaceae* from this without being able to verbalize it. Also consider very carefully what place taxonomy really holds in education, as opposed to training and learning a technique. The Victorians believed that once you had named an animal or plant, this somehow gave you power over it, and also elevated you above those who did not know the name – this attitude, used by Apache Indians and occurring in the Rumpelstiltskin story, has its roots deep in folklore and mythology, but is it education?

4 Add your own experiences.

Uses
What concepts can be established through field work?

Simplest level	*Concept*
1 *What is there?* Kids really do like looking into ponds and streams, rock pools and under stones	Taxonomy
2 *Why is it there?* (rather than somewhere else) Will it survive there?	Variation within a species Adaptation, Colonization Succession, Survival values Seasonal changes
3 *Why is it* where *it is?*	Zonation, not only in seaweeds but in hedges, tall trees, forests, paths
4 *How many?*	Population variables and controls – you don't necessarily need Fisher,[37] Ford[38] and Andrewartha[3]

Biology's special needs 133

5 *What is helping it or hindering it?* Interaction, including Man
6 *How is it recycled?* Death seen as a continuing part of the process
Energy flow – this is perhaps simpler in urban than in other environments

Approach

Too often the approach is rather head-on, seeing what has happened and using the behavioural objectives of accurate observation and description rather than those of experimental design and hypothesis formation, which in this case are much more fun and help greatly in the retention of an otherwise undigested mass of facts.

For example, colonization can be investigated by clearing a stretch of ground (incidentally sieving out all the seeds, listing all plants there in the first place and then watching colonization occur). Alternatively prepare a sterile environment such as acid soil, sand or a series of tanks as in Revised Nuffield Biology (*Teacher's Guide,* Text 3)[79] and watch what takes place. The disadvantages are those inherent in all long-term experiments – window-cleaners drop detergent in the tanks, weed killer drifts over your cleared ground, and worst of all, motivation is lost by all but the most conformist. A film record can help.

Similarly, taxonomy if met head-on can be off-putting. It is better here again to use a problem-solving approach – how is this thing making life easy for itself? Great discussion on very simple hypotheses about the use of appendages, of seed dispersal methods, of vegetative reproduction, such as in nettle and dandelion.

Examination of an artificial situation often teaches classes how to look at a natural one – and one of the greatest problems in field work is that of perception – most of the conclusions are several stages remote from the observations. There are, for example, trout farms – as at Nailsworth, near Stroud. Once you know what the fish farmer does, you look with sharpened observation at the natural habitat of fish and ask some relevant questions.

Small, familiar, local beginning

School paths Moss and weed colonization – where and why? Map-making gets a good start, so does interaction, and taxonomy is at a low premium unless you specialized in mosses! Simple names and above all descriptions of the relatively few species are all that is necessary.

In a town there may not be much moss on school yards, but there will almost certainly be some nearby – it needs you to go and find it. You might try to get your class to design a nature trail. Trails have been successfully made for such unlikely situations as Baker Street in central London.

Xeromorphs Great motivators, heaven knows why – Western movies perhaps – and soon the lab gets filled with cacti. Gives a good appreciation of one specialized habitat and a good start to informal field work – how many xeromorphs on Clapham junction? But, goodness, there are ferns growing out of the wall as well. And so to microhabitats.

One Tree What lives in, on and under it? What kind of leaf mosaic has it? Why? What protection does it need in a school or an urban environment? Forestry Commission publications from HSMO[33, 34] are useful and well produced – they also have a good weed identification booklet[39] so that you can identify the weeds from their seedlings, with a bit of luck.

Old gardens behind blocks of flats are often containers of old fruit trees with every known parasite from woolly aphids to charming little wasps. In Bradford kids collected over 700 different species of insects, and that was before the air pollution Acts! You can study leaf litter, even in cities, and car radiators are good collectors of other specimens.

Building and demolition sites   A fairly recent example was the Nine Elms site, now the new Covent Garden Market. Totally denuded and then recolonized, it is now, alas, built on. For ideas of what can be done to a derelict site, see Jeremy Cotton's account[28] of the ecology park in the centre of Southwark, just over the river from the Tower of London. It's also worth checking on the local bye-laws about the replacement of topsoil, indiscriminate tipping and so on.

Birds School playgrounds are full of them – try pigeons, variation, courtship, different types of food they will take, diurnal movements by starlings from suburbs to the city centre, gull species, flight movements. Interaction with man – pigeon damage and anti-pigeon devices.

Rivers Even in cities, where there is a river there may be accessible areas for study. Take full safety precautions and check on the access routes carefully. Mud flats can be full of worms and even Foraminifera. You may be able to get reports on the organisms found trapped on the filters of the water cooling inlets of riverside power stations; this can lead to considering pollution levels. The organisms present in the river can be used to provide a biological index of pollution levels.[8, 136] This, in turn, leads on to sewage and water purification plants. Reservoirs, such as those at Staines, near London, are good bird sanctuaries, yet accessible from the city.

Cemeteries For the committed dialectical materialists try the ecology around Karl Marx's tomb in Highgate cemetery, if it is open and if you can shoulder your way through the visitors! But seriously, cemeteries have considerable potential for simple field work. Air pollution may erode tombstones and expose fossils. In turn eroded stone provides a substrate for moss colonization and so on. Tombstones are also good habitats for lichens – and they are excellent biological indicators of air pollution.[7,116] In this way the class can make simple measures of pollution levels and possibly compare them with those obtained by chemical and physical means.

So the message is – look for potential in your own surroundings. Field work is too important to be left solely to one week's stint at a field centre for the Lower Sixth. Of course that may be, and often is, an invaluable experience for them. But that's only a very small proportion of your biology pupils, and field work is a valuable experience for all. So what do you intend to do about it?

Appendix A An example of the historical approach

New teachers accustomed to the use of history as an *introduction* to a piece of work, are warned that at school level, the historical approach confuses new ideas rather than illuminates them. It is therefore recommended as a method of *consolidation,* especially for fast finishers; as it was used in the following worksheet. The description of Malpighi's experiments is adapted from a review in *Scientific American.**

Malpighi (1628–94) was a professor of medicine at the University of Bologna in Italy. He was very famous for his biological discoveries. We still remember him today because part of the kidney is named after him (see if you can find out which part).

A friend of his, Professor Borelli, lived in another part of Italy and the two wrote to each other about their discoveries. Borelli suggested that his friend should write a couple of articles about his study of the lungs.

At that time William Harvey had already discovered the circulation of the blood. It was known that there were arteries and veins and that the blood moved away from the heart in one and toward the heart in the other. The nature of the connection between them, however, was still mysterious; *that is, the capillaries were unknown.* Similarly, it was known that air enters the lung through the trachea, but neither the structure of the lung nor its function was understood. One can appreciate Malpighi's problem. Here are this mass of tissue, the lungs. They swelled with air like a sponge. Blood entered the lungs from the heart and returned to the heart from the lungs. What were they? What did they do? How did they do it?

Malpighi's first experiments presented him with a contradiction. A cat's lungs were removed and the trachea was pinched shut; when the lungs were squeezed, air did not come out of the severed blood vessels, as it would have if the trachea were connected to the blood vessels.

In another experiment the lungs were filled with water through the trachea. When they were squeezed some of the water appeared in the left ventricle, having got there through the pulmonary vein. The second experiment supports the notion that the arteries, the veins and the trachea are connected; the first contradicts it. The breakdown of capillaries that must have occurred in the second experiment was to confuse Malpighi almost to the end of his researches on the lung.

---

* Maxwell H. Braverman: review of Howard E. Adelmann, *Marcello Malpighi and the Evolution of Embryology* (Cornell University Press) in *Scientific American,* vol. 216, no. 4 (1967).

Appendix A

Next, looking at the same question in another way, he injected water coloured black into the pulmonary artery of a sheep and observed that the sheep's lungs swelled. The outside of the lungs was blackened, but he does not describe water coming out of the trachea. It should have done if there had been a connection between the blood vessels and the trachea.

Professor Borelli replied to his friend's letters about the lung experiments. He was impatient with the wrong ideas that Malpighi had. Malpighi still believed that the pulmonary artery, the pulmonary vein and the trachea were all joined together.

Malpighi's observations led him to conclude that, contrary to what was then thought, the lungs were not just flesh but consisted of inflatable thin bubble-like structures. He published this conclusion in his first article on the lungs. He added that there was free entry to these structures from the trachea. He said that it could be demonstrated by removing the lungs from a frog, washing the blood out of them, injecting water into the pulmonary artery, inflating the lungs by blowing into the trachea and tying them up so that they remained inflated as they dried. Having done this, he said, one could observe the extremely thin membranes and also the wonderful network of vessels around them, *which he supposed to be nerves*. He further described the branches of the trachea, and of the pulmonary artery and vein to their finest visible subdivisions. But he left unsettled the relationships between the three vessels. In closing he suggested that the function of the lungs was to mix the liquid and solid parts of the blood and so cause their liquefaction.

After Borelli had read about this he wrote back to Malpighi 'The more I think about this mixing, the greater the difficulty I find with it. . . .I should like you to investigate further how bodies can be kept fluid'.

1 Referring to your diagram of the blood vessels round the alveolus, say what happened in each of the four experiments.
2 Explain the reasons for Malpighi's mistakes.
3 Now explain what he did and thought to a member of your group who has not had time to look at this worksheet.

Appendix B Worksheets used in team-teaching

Part 1 WHAT IS NORMAL?

Have a look at yourself and try to find the answers to these questions.

YOUR WHOLE BODY

a) How tall are you without shoes? ...... cm

b) What is your height when you are sitting? ...... cm
EVERYONE MUST SIT IN THE SAME CHAIR? BUT NOT AT THE SAME TIME!

c) How heavy are you without shoes? THIS IS YOUR <u>MASS</u>! ...... kg

d) How much skin do you have? To answer this you have to find the <u>AREA</u> of your body surface. How????? ...... cm^2
(Ask your teacher if you don't know.)

e) How could you find your <u>VOLUME</u>?

....

Well no, I don't THINK you can do it in class, but ask your teacher!

YOUR HANDS

a) How long are your fingers and thumb?

LEFT HAND		RIGHT HAND	
Thumb	...... cm	Thumb	...... cm
First finger	...... cm	First finger	...... cm
Second finger	...... cm	Second finger	...... cm
Third finger	...... cm	Third finger	...... cm
Little finger	...... cm	Little finger	...... cm

b) What <u>area</u> does your hand cover? ...... cm^2

Worksheets used in team-teaching

c) What is the width of your hand span?

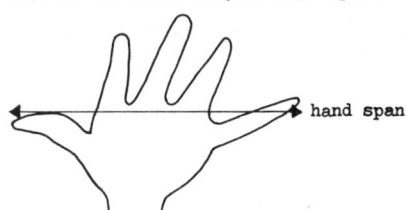

Left hand ...... cm

Right hand ...... cm

d) What is the volume of your hand?

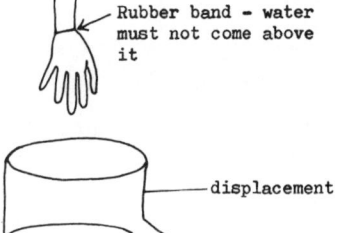

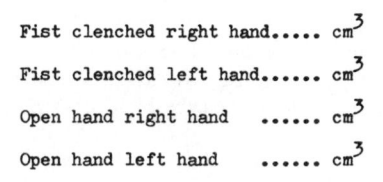

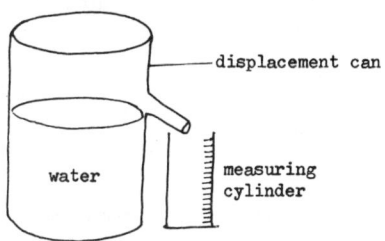

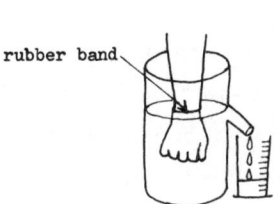

Rubber band - water must not come above it

Fist clenched right hand..... cm^3

Fist clenched left hand...... cm^3

Open hand right hand ...... cm^3

Open hand left hand ...... cm^3

displacement can

water

measuring cylinder

rubber band

e) How much can your hand pick up?

To answer this, FIND OUT how many sunflower seeds you can pick up. 10 sunflower seeds = one gram., so pick them up and plop them onto a weighing machine.

Right hand ...... g x 10 = ... seeds.

Left hand ...... g x 10 = ... seeds.

f) How strong is your grip?

Squeeze the bottle ONCE ONLY and see how far up the tubing the liquid goes.

Right hand ...... cm

Left hand ...... cm

YOUR FEET AND LEGS

a) How LONG are your feet? Well, yes, you'd better take off your shoes. Right foot ...... cm   Left foot ...... cm

b) What is the <u>area</u> of your feet? (Ask your teacher for squared paper) Right foot ...... cm^2 Left foot ...... cm^2

c) How far can you jump straight up in the air?

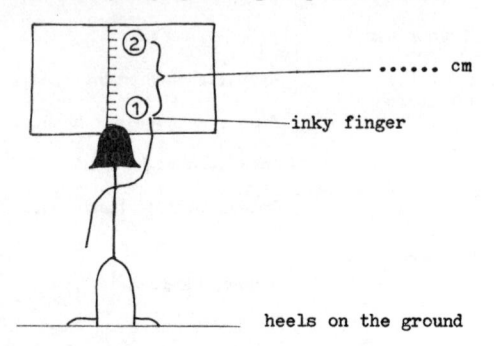

...... cm

inky finger

heels on the ground

1 First dab your fingers on the ink pad.
2 Then face the paper on the wall with your heels on the ground and your inky finger arm held as high as possible above your head.
3 Make a mark on the paper with your finger.
4 Jump as high as you can and make a second mark.
5 Measure the distance between the two marks.

d) How far can you jump ALONG from a standing position?

Stand with both feet on a line on the floor and jump along as far as you can. Measure the distance.

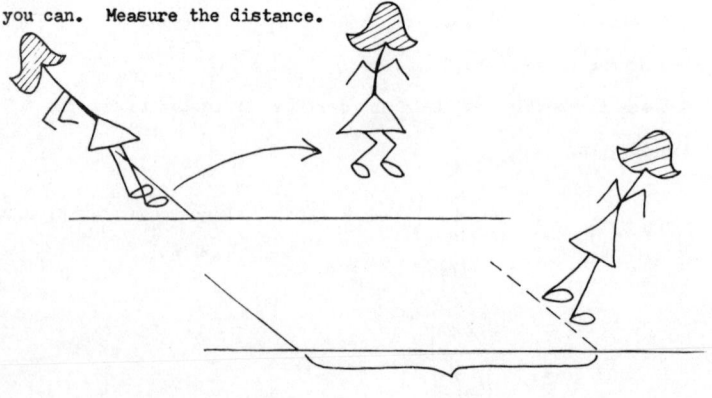

...... cm

Worksheets used in team-teaching 141

BODY ACTIVITIES

a) <u>How quick are you?</u>

Work with a partner. One person holds the marked card at the top, and the other holds a thumb and first finger at each side at the BOTTOM but without touching it.

How soon can you catch it when your partner drops it?

At what mark have you grabbed the card?

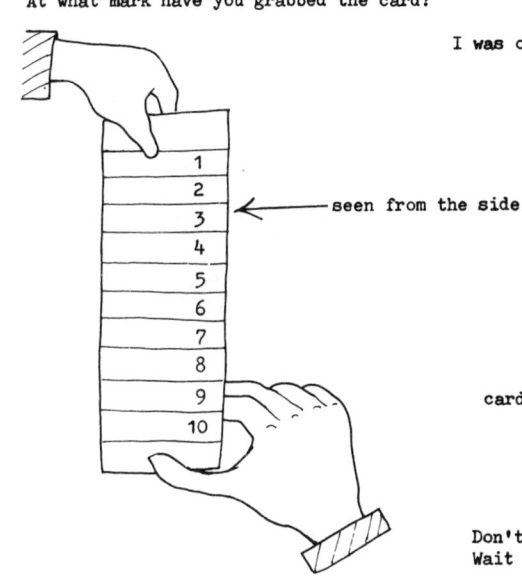

I was on mark ......

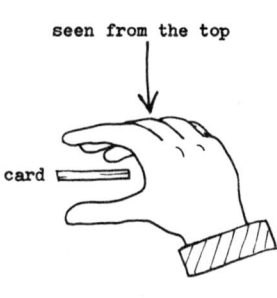

seen from the side

seen from the top

card

Don't touch the card at first! Wait until it drops.

b) How big are you round your chest?
How much bigger can you make it?
What is your chest size?   ...... cm
Breathe <u>out</u> as much air as you can -
what's your chest size now? ...... cm (a)
Breathe <u>in</u> as much air as you can -
now what is your chest size? ...... cm (b)
So how much can you make your chest
expand? (b - a) cm = ... cm

Hold up your arms while your partner puts the tape measure round.

c) Rate of breathing

   Using a stop watch, count the number of breaths you take in 60 seconds. I took ...... breaths.

d) How long did it take you to do your last maths homework?

   ...... minutes

e) How long did it take you to do your English homework?

   ...... minutes

THE END!

And about time, too!!

Oh! Are you measuring your sitting area? WHAT a good idea! Have some graph paper!

Worksheets used in team-teaching   143

Teacher's notes   WHAT IS NORMAL?

<u>Part 1</u>   Taking measurements.

Equipment needed:

YOUR WHOLE BODY

Height	Tape or metric rule fixed to wall (cm).
Mass	Weighing machine (kg).
Skin area	Newspapers, scissors and Sellotape. Rulers. Needs to be wrapped around body and then measured. A popular riot raiser in mixed class.

YOUR HAND

Length of fingers	Tape measure or paper and ruler.
Area of hand	Graph paper or squared paper (cm)
Volume	Displacement can, rubber bands for wrists and measuring cylinders.
Span	Ruler or tape measure (cm).
Amount of pick up	Sunflower seeds, weighing machine.
Strength of grip	Polythene bottle and plastic tubing fixed to wall beside metre rule.

YOUR FEET AND LEGS

Length	Ruler or tape measure.
Area	Graph paper or squared paper (cm).
Jumping up	Ink pad. Plain paper stuck on wall at appropriate height, ruler or tape measure.
Jumping along	Line on floor, ruler or tape measure.

BODY ACTIVITIES

Reaction time	Marked card, 3 cm wide, 25 cm long, marks 1-10 equidistant.
Chest size	Tape measure.
Rate of breathing	Stop watch, watch or clock with second hand.

Part **11**   WHAT IS NORMAL?

YOUR WHOLE BODY

a) 1. What is the most common HEIGHT in the class?   ...... cm
    2. How tall is the tallest member of the class?   ...... cm
        How tall is the shortest member of the class?   ...... cm
    3. Try to find out the height of your mother.   ...... cm
        Try to find out the height of your father.   ...... cm
    4. Do short parents seem to have short children?   YES or NO
        Do tall parents seem to have tall children?   YES or NO

b) What is the shortest sitting height in your class?   ...... cm
    What is the tallest sitting height in your class?   ...... cm
    Are these the same people as the ones who are the
    tallest and shortest when standing?   YES or NO
    If the answer is NO, give some reasons why they
    are not the same people.

c) 1. What is the mass of the heaviest member of the
        class?   ...... kg
        What is the mass of the lightest member of the
        class?   ...... kg
    2. Are the tallest people usually the heaviest and
        the shortest people the lightest?   YES or NO

YOUR HANDS

a) 1. Does the smallest hand span cover the smallest area?   ......
    2. How useful is a span as a unit of measurement?   ....

b) What difference, if any, does it make to your measurement of the
    volume of your hand if you have your fist open or closed? ....
    ....

c) Find the hand in the class which picks up the largest number of
    seeds and the one which picks up the smallest number of seeds.

1. In what ways are the two hands different?

Feature	Most seeds hand	least seeds hand
length of fingers size of span what else?		

2. How would you answer the question <u>'How big is your hand?'</u>
....

d) 1. Which hand do you use for writing?   Right/Left

    2. Is your writing hand your strongest?   Yes/No

YOUR FEET AND LEGS

a) What sort of person jumps up highest into the air?
Is it the tallest or who? ....
....

b) Find the person in your class who jumped ALONG the longest distance. Why do you think he/she was able to jump further than the rest of you? ....

BODY ACTIVITIES

a) 1. What might cause you to change your rate of breathing? ....
....

    2. What measurement might be connected with your rate of breathing? Plot a scattergram for your whole class and see if you were right. Ask your teacher if you are not sure what to do.

b) If a person does maths homework quickly, do they also do English homework quickly? ....

The LAST question! (True!)

Now! Try and write in ONE SENTENCE an answer to the question WHAT IS 'NORMAL'??? ....
....
....

Teachers' Notes　　　　　　　WHAT IS NORMAL?

Part <u>11</u> Why did we take all those measurements?

This is the most important part of the whole exercise as it consolidates the information collected in Part <u>1</u> and provides the means for introducing concepts such as range, mean, average, median, without necessarily using the terms.

1. <u>Using histograms</u>　　　　　　　　Suggested for:

```
No. of
people      Heights of Form 1
 16 |     |      |      |      |                    Standing height
 15 |     |      |      |      |                    Sitting height
 14 |     |      |      |      |                    Mass
 13 |     |      |      |      | ← Ruled            Area of hand
 12 |     |      |      |      |   guidelines       Hand span
 11 |     |      |      |      |                    Amount picked up by hand
 10 |     |      |      |      |                    Length of feet
  9 |     |      |      |      |                    Area of feet
  8 |     |      |      |      |                    How far can you jump up?
  7 |     |      |      |      |
  6 |     |      |      |      |                    How far can you jump along?
  5 |     | Leroy|      |      | Coloured sticky
  4 |Bill | Jill | Jean |      | paper - big        Reaction time
  3 |Ted  | Sue  | Abul |      | enough to write    Chest size
  2 |Jerry| Ian  | Raza |      | names on           Chest expansion
  1 |Jane | Jo   | Jim  | John |                    Time to do homework
              Height(cm)
```

Prepare these sheets in advance - keep to compare with next year's first year. No, they don't have to plot them all! Four parameters are enough.

2. <u>Using scattergrams</u>: for correlation.

Suggestions for scattergrams:-
Standing height/Jumping up or along
Standing height/Mass
Area of hand/Hand span
Hand span/Amount picked up
Rate of breathing/Chest expansion
Reaction time/Homework time

These are often more popular - the class might want to do them all!

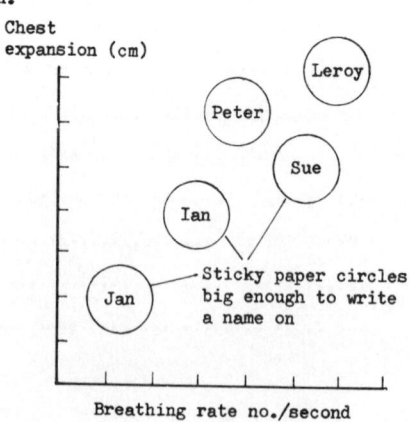

Chest expansion (cm)

(scattergram showing Leroy, Peter, Sue, Ian, Jan plotted against Breathing rate no./second, with note "Sticky paper circles big enough to write a name on")

Appendix C   Use of the clinical thermometer

THE USE OF THE CLINICAL THERMOMETER

> Knowledge assumed: of thermometers and lenses.
>
> Apparatus required: one clinical thermometer for each pupil (stubby bulb)
> Graphs for display of class result
> Diluted hypochlorite solution
>
> Method: preferably with a whole class. This ensures a good distribution curve; results of parallel classes may be added.
>
> Shake the thermometers before use, to ensure that the reason for shaking is discovered.

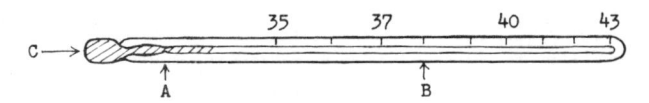

LOOK hard at your thermometer. What do you think the narrow part A is for?

Is there any good reason why the glass is so thick at B?

Why is the glass different again at C?

RINSE the thermometer in the disinfectant.

Put end C under your tongue and close your mouth. Leave it there for TWO minutes. Then take it out and read your temperature. Now shake it very sharply. Take your temperature again. Is it any different? Why do you think shaking is important?

OTHER places to take your temperature include under your arm. Is your temperature the same there as it is under your tongue?

MARK your temperature on the class graphs. What is the temperature of most of the class? What else do the class graphs show that your own two results did not?

TAKE your temperature under your tongue:

1. After exercise.
2. After drinking hot tea.
3. After breathing through your mouth for five minutes.

Can you think of any other things which could change your temperature? Investigate them. Does everyone's temperature change the same amount for the same amount of exercise?

Appendix C

<u>CAREFULLY</u> TAKE YOUR THERMOMETER HOME. Take your temperature first thing in the morning:

    BEFORE you get out of bed
    BEFORE you have a hot cup of tea
    BEFORE you take deep breaths of cold air.

Mark your readings on this graph — and try to remember to do it for 35 days.

Temperature	NUMBER OF DAYS (1–35)
37.5	
37.25	
37.0	
36.9	
36.75	
36.5	
36.25	

When you have done this, bring your graphs to school and trace them on transparent film. Sort them out into groups which are the same. Can you notice any differences between boys and girls? Fat and thin people? People of different ages? What else do the graphs tell you?

<u>YOU MIGHT ALSO LIKE TO</u>:
graph a baby's temperature or those of your parents and grandparents.
graph your temperature at teatime and last thing at night as well.

INFORMATION FOR TEACHERS

Death occurs if the body temperature goes above $44°C$ or below $21°C$. Body temperature lies between 36.5 and 37.5 degrees Celsius. Variations:- temperatures taken in the groin are usually the same as those taken in the mouth; armpit temperatures usually one degree lower and rectal temperatures about one degree higher.

<u>Lowering of temperature occurs</u> with emotion, due to tensing muscles, with old age, with mouth breathing (people with blocked nostrils due to a cold may therefore not have the high temperature in the mouth that they can feel in the rest of their body), after cold drinks (say down to $35°$ for about 30 minutes), with the lessening of physical activity as night approaches (say $35.5°$ at midnight), with a cold bath ($36.5°$) and after menstruation.

<u>Raising of temperature occurs</u> possibly due to the time of day ($37.25°$), with athletic exercise (say $40.5°$ - although skin temperature will be reduced at the same time due to sweating), and after a hot bath ($37.75°$), a hot drink ($37.75°$) and after ovulation ($37.5°$).

<u>The life pattern of temperature changes</u> - the newborn baby has a very poor temperature regulating system and can drop $1.5°$ in temperature for each of the three hours directly after birth. At about 18 months they are at their 'hottest' averaging $37.25°$; this gradually drops to $36.9°$ when girls are 14 and when boys reach 18. The cult of temperature taking in children is therefore meaningless, as alarming fluctuations are the norm rather than the exception; nor does one need a thermometer to confirm a child's high fever!

Labour wards are well-heated to help the baby's chances of survival. Operating theatres similarly because anaesthetics reduce the body's ability to maintain $36.9°$ - thus working conditions for surgeons are similar in temperature to those of coalminers, and each need extra salt, as sweating can be excessive.

Appendix C

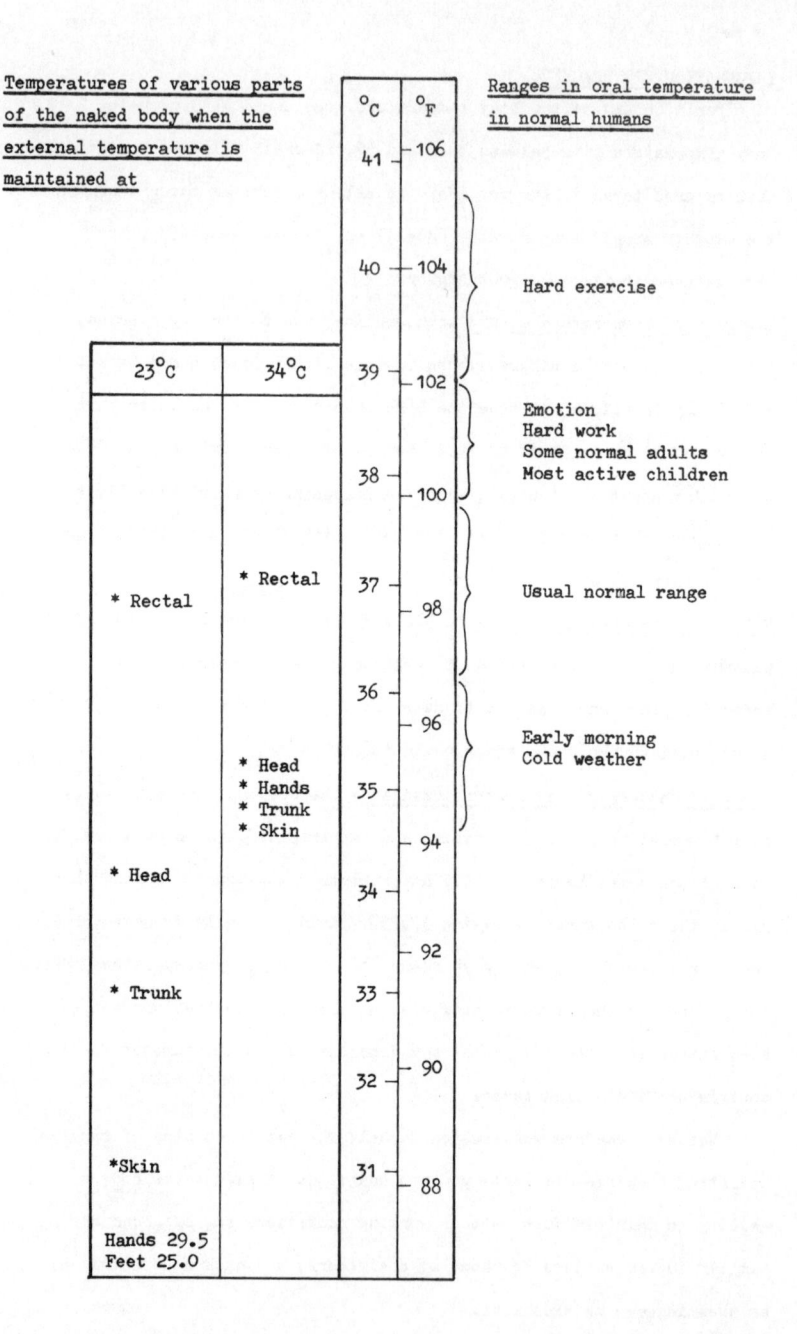

Temperatures of various parts of the naked body when the external temperature is maintained at

23°C	34°C
* Rectal	* Rectal
	* Head * Hands * Trunk * Skin
* Head	
* Trunk	
*Skin	
Hands 29.5 Feet 25.0	

°C	°F
41	106
40	104
39	102
38	100
37	98
36	96
35	94
34	92
33	90
32	88
31	

Ranges in oral temperature in normal humans

} Hard exercise

} Emotion
Hard work
Some normal adults
Most active children

} Usual normal range

} Early morning
Cold weather

Appendix D *Quest* worksheets: basic skills

QUEST   00101 Q

Under each of the pictures write one of the 3 words (hot, cold, warm) which fits best.

152 Appendix D

QUEST 00103 Q

Cut up this page. Stick the parts in your book.

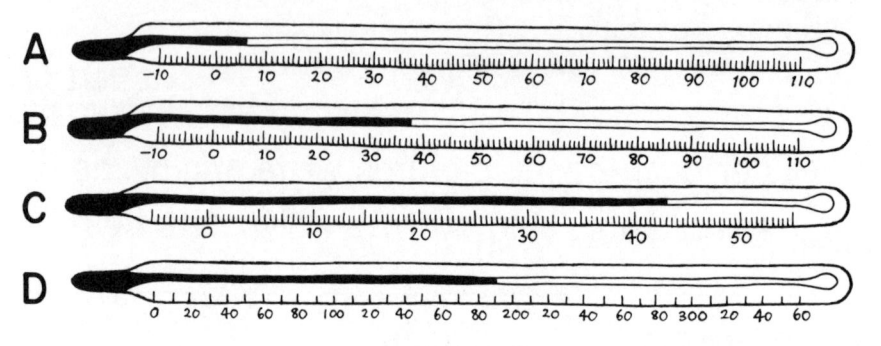

A
B
C
D

The temperature shown by thermometer –

A is _____ B is _____ C is _____ D is _____

Mark the thermometers for these readings

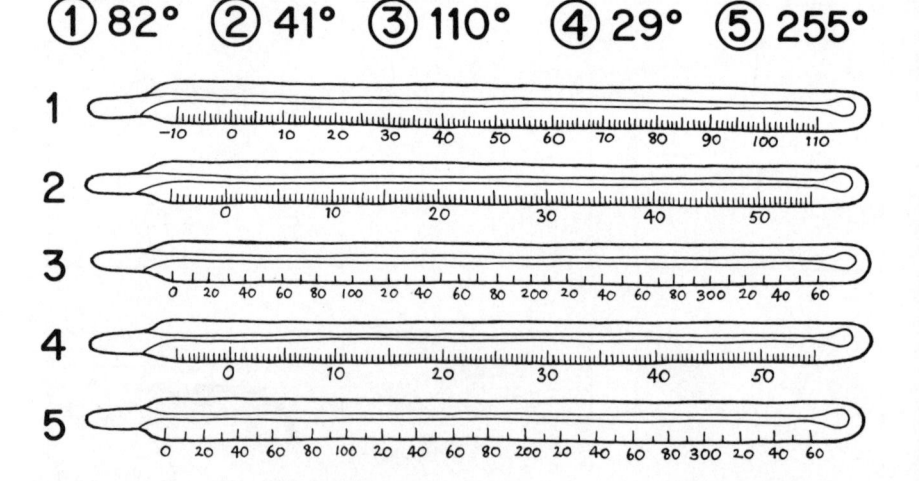

① 82°   ② 41°   ③ 110°   ④ 29°   ⑤ 255°

1
2
3
4
5

Appendix E   An independent learning approach for the sixth form

DIGESTION

Make sure that you understand tooth structure and function as adaptations to diet.

What causes flow of saliva in the mouth? ....
Is this under nervous or humoral control? ....
What is the enzyme present in saliva and its action? ....
Under what pH does it work best? ....

```
     bolus ─────                    ────── circular muscle
```

This diagram shows the passage of a bolus of food along the oesophagus.
What is this type of movement called? ....
Describe how the activity of the circular muscles brings about this movement. ....
....
In which direction will the bolus of food shown move? ....
This layer of circular muscle runs the whole length of the gut, though it is thicker in some parts than in others. In three places it is particularly thick and forms valves. What are the names of these valves and where are they? ....
....
What is their function? ....
What are the enzymes secreted by the stomach? ....
From which glands and from which part of the stomach are they secreted?
....
What do they act upon and under what pH? ....
What is necessary to acquire this pH? ....
Pepsin is secreted in the inactive form of pepsinogen. Pepsin is an endopeptidase. What does this mean? ....
....

Apart from digestion name 3 other functions of the stomach. ....
....
....

NEED FOR CONTROL AND SYNCHRONISATION OF DIGESTIVE JUICES

This is necessary so that the enzymes are only secreted when the appropriate food is present, to ensure their economic expenditure. They must be secreted in the appropriate sequence and for the right length of time. This co-ordination is brought about by the actions of nerves and hormones.

Pavlov performed experiments on dogs to find out what controls gastric juice secretion. He disconnected the oesophagus from the stomach, so that food swallowed never reached the stomach. He fed the dog and collected its gastric juice. Five minutes after feeding there followed a copious highly acid gastric secretion which lasted for 3 to 4 hours but gradually diminished.

He then cut the vagus nerve (this is the Xth cranial nerve innervating the viscera). <u>Result</u>: no gastric juice secreted.

In a normal dog the vagus nerve was stimulated but no food given. <u>Result</u>: copious gastric juice secreted.

1. What stimulates the release of gastric juice? ....
   This initial copious flow of gastric juices is called appetite juice. After a meal containing much protein the hormone gastrin is found in large amounts in the blood.
   If food is given to a dog with a severed vagus nerve, there is no change in the amount of gastric juice secretion but the juice contains no pepsin.
2. What does this tell you about the control of gastric juice secretion? ....
   Which mechanism would you think starts the flow of gastric juice?
   ....
   Which mechanism maintains it? ....
   This is analogous to starting a car by using a battery. Explain in terms of this analogy. ....
   What would you do next?   ....

Appendix F Resource lists

Before you use these lists, consider the following points:

1 They are not intended to be, and in fact cannot be, complete. There are always new firms appearing and new materials from old. You have to keep an eye open for changes by scanning journals such as *School Science Review* and *Journal of Biological Education*.
2 Firms move, shut down, telephone numbers and even post codes change. Any list is to some extent out of date before it is published. So don't get too frustrated when you can't locate a listing.

1 Sources of audio-visual aids for biology teaching

Some non-commercial audio-visual firms, such as multinational companies, market materials at cost, or even free. This also applies to hiring 16 mm films from a few film libraries.

Remember that if you are ordering a 16 mm film you will have to book it at least a month ahead – and perhaps even three or four months, due to the demand. This may make the problem of integrating it into your teaching a difficult one.

Adam Rouilly & Co. Ltd, Crown Quay Lane, Sittingbourne, Kent ME10 3JG, tel. [0795] 71 378. *Anatomical models and charts*
Argus Communications, Edinburgh Way, Harlow, Essex CM20 2HL, tel. [0279] 39441/4. *Posters with catchy phrases*
Arnold, E. J., Butterley Street, Leeds LS10 1AX, tel. [0532] 442 944. *35 mm slides, film strips, overhead projector transparencies*
Audio Learning, Sarda House, 183-189 Queensway, London W2 5HL, tel. 01-727 2748. *Audio tapes with 35 mm slides*
Audio-Visual Productions, Hocker Hill House, Chepstow, Gwent NP6 5ER, tel. [029 12] 5439. *35 mm slides, overhead projector transparencies*

Banta (Educational Suppliers) Ltd, 279 Church Road, London SE19 2QQ, tel. 01-653 4798. *35 mm slides, 'Bioviewers' and 'Biosets'*

Basil Blackwell & Mott Ltd, 108 Cowley Road, Oxford OX4 1JF, tel. [0865] 724041. *Plant ecology charts (British Museum series)*

BBC Publications, 35 Marylebone High Street, London W1M 4AA, tel. 01-580 5577. *Booklets, film strips, film loops, records*

BBC Television Enterprises, Film Hire Library, Woodston House, Oundle Road, Peterborough, Northants PE2 9PZ, tel. [0733] 52 257/8. *16 mm films from BBC programmes*

Boulton-Hawker Films Ltd, Hadleigh, Ipswich, Suffolk IP7 5BG, tel. [047 338] 2235. *Films for sale only. Hire from NAVAL*

BP Educational Service, PO Box 5, Wetherby, West Yorks LS23 7EH, tel. [0937] 843 477. *'Ridpest File', a biological game*

BP Film Library, 15 Beaconsfield Road, London NW10 2LE, tel. 01-451 1129. *16 mm films, conservation, ecology*

British Film Institute, 127–133 Charing Cross Road, London WC2H OEA, tel. 01-437 4355. Film hire from: British Film Institute, Distribution Library, 42–43 Lower Marsh, London SE1 7RG, tel. 01-928 4743. *16 mm films, wide range*

British Gas Film Library, Park Hall Road Training Estate, London SE21 8EL, tel. 01-670 6161. *16 mm films, gas, air pollution*

British Life Assurance Trust for Health Education, BMA House, Tavistock Square, London WC1H 9JP, tel. 01-388 7976. *Medical audio-visual programmes*

British Museum (Natural History), Publications Officer, Cromwell Road, London SW7 5BD, tel. 01-589 6323. *Charts on ecology (personal collection), postcards, models*

British Transport Film Library, Melbury House, Melbury Terrace, London NW1 6LP, tel. 01-262 3232. *16 mm films, natural history*

Brookwick, Ward & Co. Ltd, 8 Shepherd's Bush Road, London W6 7PQ, tel. 01-743 1847. *Anatomical models*

Cambridge University Press, 200 Euston Road, London NW1 2DB, tel. 01-387 5030. *Film strips, tapes, games for* Basic Biology Course *(order through booksellers)*

Camera Talks Ltd, 31 North Row, London W1R 2EN, tel. 01-493 2761. *Film strips, film loops, tape commentaries*

Cancer Information Association, Gloucester Green, Oxford OX1 2EQ, tel. [0865] 46 654. *Film strips, booklets*

Capital Biotechnic Developments, 66A Churchfield Road, London W3 6DL, tel. 01-992 5824. *Molecular models*

Central Electricity Generating Board, Film Library, Press and Publicity Office, Sudbury House, 15 Newgate Street, London EC1A 7AU, tel. 01-248 1202. *16 mm films, conservation*

Central Film Library, Government Building, Bromyard Avenue, London W3 7JB, tel. 01-743 5555. *16 mm films*

Centre for World Development Education, Parnell House, 25 Wilton Road, London SW1V 1JS, tel. 01-828 7611. *Charts, pamphlets, film strips, list of resource materials*

Cochranes of Oxford Ltd, Leafield, Oxford OX8 5NT, tel. [099 387] 641. *Molecular models*

Concord Films Council, 201 Felixstowe Road, Ipswich, Suffolk IP3 9BF, tel. [0473] 76 012 *16 mm films, Third World, social issues, conservation.*

Council for Nature, c/o Zoological Society of London, Regent's Park, London NW1 4RY, tel. 01-722 7111. *16 mm films, natural history, conservation and list of such films*

Crystal Structures Ltd, Bottisham, Cambridge CB5 9EA, tel. [0223] 81145. *Molecular models*

Diana Wyllie Ltd, 3 Park Road, Baker Street, London NW1 6XP, tel. 01-723 7333. *Film strips and tape commentaries*

Dundee College of Education, Co-ordinator of Learning Resources, Gardyne Road, Broughty Ferry, Dundee DD5 1NY, tel. [0382] 453 433. *Self-instructional multi-media materials*

Educational & Scientific Plastics, 76 Holmethorpe Avenue, Holmethorpe, Redhill, Surrey RH1 2PF, tel. Redhill 62787/8. *Plastic anatomical models*

Educational Productions, Bradford Road, East Ardsley, Wakefield, West Yorks WF3 2JN, tel. [0924] 823971. *Film strips, charts, overhead projector transparencies, game 'Man in his Environment', etc.*

Edward Arnold (Publishers) Ltd, 41 Bedford Square, London WC1B 3DQ, tel. 01-637 7161. *Computer simulation materials (order through bookseller)*

Edward Patterson Associates, 68 Copers Cope Road, Beckenham, Kent BR3 1RJ, tel. 01-658 1515. *Multimedia materials on various topics*

Eothen Films Ltd, EMI Film Studios, Shenley Road, Borehamwood, Herts WD6 1JG, tel. 01-953 1600. *16 mm films, film loops, 35 mm slides*

Fact and Faith Films, 37 Coton Road, Nuneaton, Warwickshire CV11 5TW, tel. [0682] 381 690. *16 mm films*

Fauna Preservation Society, c/o Zoological Society of London, Regent's Park, London NW1 4RY, tel. 01-586 0872. *16 mm films, conservation of wildlife*

Fergus Davidson Associates Ltd, 376 London Road, West Croydon, Surrey CRO 2SU, tel. 01-689 6824. *16 mm films (sale only) film strips, study prints, overhead projector transparencies* (Encyclopedia Britannica)

Focal Point Audio Visual Ltd, 35 Cavendish Drive, Waterlooville, Portsmouth, Hants PO7 7PJ, tel. [070 14] 3111. *Film strips*

Foundation for Teaching Aids at Low Cost (TALC), Institute of Child Health, 30 Guildford Street, London WC1N 1EH, tel. 01-248 9789. *35 mm slides, flannelgraphs on medical, nutritional and other areas for overseas use*

Frederick Warne Co. Ltd, 40 Bedford Square, London WC1B 3HE, tel. 01-580 9622. *Charts on natural history*

French Scientific Film Library, c/o Contemporary Films, 55 Greek Street, London W1V 6DB, tel. 01-734 4901. *16 mm films, scientific*

Gateway Educational Media, Waverley Road, Yate, Bristol BS17 5RB, tel. [0454] 316 774. *Film strips and loops, films.* Hire from: Gateway Film Hire Library, 15 Beaconsfield Road, London NW10 2LE, tel. 01-451 1127. *16 mm films*

General Dental Council, 37 Wimpole Street, London W1M 8DQ, tel. 01-486 2171. *Various visual aids about teeth*

German Federal Republic, Embassy of, Film Library, 67 Great Chapel Street, London WC1A 2SL, tel. 01-734 9102. *16 mm films, Germany, natural history*

Gerrard, T. & Co., Gerrard House, Worthing Road, East Preston, West Sussex BN16 1AS, tel. [090 62] 72071/5. *Film strips, models, charts, overhead projector transparencies, etc.*

Guild Sound and Vision, Woodston House, Oundle Road, Peterborough, Northants PE2 9PZ, tel. [0733] 63 122. *16 mm films*

Harcourt Brace Jovanovich, 24–28 Oval Road, London NW1 7DX, tel. 01-485 7074 *BSCS inquiry slide sets*

Health Education Audio Visual, Neatham Mill, Lower Neatham Lane, Holybourne, Alton, Hants GU34 4ET. *Health education film strips, tapes, etc.*

Health Education Council, 78 New Oxford Street, London WC1A 1AH, tel. 01-637 1881. *Leaflets, posters, display materials, resource lists, etc.*

Hodder & Stoughton Ltd, Mill Road, Dunton Green, Sevenoaks, Kent TN13 2XX, tel. [0732] 50 111. *Charts on handling and sexing of small mammals*

Hugh Baddeley Productions, 8 Brampton Road, St. Albans, Herts, AL1 4PW, tel. Potters Bar 54046. *Film strips*

Hulton Educational Publications Ltd, Raans Road, Amersham, Bucks HP6 6JJ, tel. [024 03] 4196.

ICI Film Library, 15 Beaconsfield Road, London NW10 2LE, tel. 01-451 2992. *16 mm films, agriculture, nature*

ITL Vufoils Ltd, 10–18 Clifton Street, London EC2A 4BT, tel. 01-247 7305/8. *Overhead projector transparencies*

John Murray (Publishers) Ltd, 50 Albemarle Street, London W1X 4BD, tel. 01-493 4361. *35 mm slides, film loops, overhead projector transparencies*

Liberation Films, 2 Chichele Road, London NW2 3DA, tel. 01-450 7850/6. *16 mm films, 'Trigger' films on aspects of health education*

Longman Group Ltd, Pinnacles, Harlow, Essex CM19 5AA, tel. [0279] 29 655. *35 mm slides, film strips, film loops*
Longman Group, Resources Unit, 35 Tanner Row, York YO1 1JP, tel. [0904] 20 801. *Science games, multimedia materials*

Macmillan Co. Ltd, Houndsmills, Basingstoke, Hants RG21 2XS, tel. [0256] 29 242. *Film loops, charts*
Manchester Regional Committee for Cancer Education, Kinnaird Road, Manchester M20 9QL, tel. 061-434 7721. *Film strips, flannelgraph, posters, etc.*
Marian Ray, 36 Villier's Avenue, Surbiton, Surrey KT5 8BD, tel. 01-390 1800. *Film strips*

National Audio Visual Aids Library, Paxton Place, Gipsy Road, London SE27 9SR, tel. 01-670 4247. *16 mm films and film library publications on visual aids, 35 mm slides, film strips, film loops, wide range*
National Coal Board, Hobart House, Grosvenor Place, London SW1X 7AE, tel. 01-235 2020. *16 mm film, coal formation*
National Dairy Council, 5-7 John Princes Street, London W1M OAP, tel. 01-499 7822. *Charts, pictures, booklets etc.*
Nicholas Hunter Filmstrips, Mutton Yard, 46 Richmond Road, Oxford OX1 2JT, tel. [0865] 52 678. *35 mm slides*

Open University Educational Enterprises Ltd, 12 Cofferidge Close, Stony Stratford, Milton Keynes MK11 1BY, tel. [0908] 566 744. *Audio tapes*
Open University Film Library, Woodston House, Oundle Road, Peterborough, Northants PE2 9PZ, tel. [0733] 63 122. *16 mm films*
Oral Hygiene Service, Hesketh House, Portman Square, London W1A 1DY, tel. 01-486 1200. *Posters, booklets on teeth*
Oxfam, 274 Banbury Road, Oxford OX2 7DZ, tel. [0865] 56 777. *Pamphlets, charts, photographs, details of projects for schools, etc.*
Oxford Educational Resources, Audio Visual Division, 197 Botley Road, Oxford OX2 OHE, tel. [0865] 49 988. *35 mm slides by Oxford Scientific Films*
Oxford Scientific Films Ltd, Long Hanborough, Oxford OX7 2LD, tel. [0993] 881 881 *16 mm films, natural history*

Pedigree Petfoods, Stanhope House, Stanhope Place, London W2 2HH, tel. 01-723 3444. *Posters, leaflets and other information about keeping pets*
Pergamon Press Ltd, Headington Hill Hall, Oxford OX3 OBW, tel. [0865] 64 881. *35 mm slides*
Philip Darvill Associates, 280 Chartridge Lane, Chesham, Bucks HP5 2SG, tel. [024 05] 3643. *16 mm film on lice and one on pesticides.*

Philip Harris Biological Ltd, Oldmixon, Weston-super-Mare, Avon BS24 9BJ, tel. [0934] 413 063. *35 mm slides, charts, film strips, film loops, models, etc.*
Pictorial Charts Educational Trust, 27 Kirchen Road, London W13 OUD, tel. 01-567 5343 and 9206. *Charts*

Rank Audio Visual Ltd, PO Box 70, Great West Road, Brentford, Middx TW8 9HR, tel. 01-568 9222. *16 mm films, film strips, film loops*
Rentokil Ltd, Film Library, Webber Road, Kirkby Industrial Estate, Liverpool L33 7SR, tel. 051-548 5050. *16 mm films, pest control*
Rentokil Ltd, Stationery Department, Selcourt, East Grinstead, West Sussex RH19 2JY, tel. [0342] 833 022. *Charts on pests: rats, fungi, etc.*
Royal Society for the Prevention of Cruelty to Animals, The Manor House, The Causeway, Horsham, Sussex RH12 1HG, tel. [0403] 64 181. *16 mm films, 35 mm slides, charts, etc. on animal welfare*
Royal Society for the Protection of Birds, The Lodge, Sandy, Bedfordshire SG19 2DL, tel. [0767] 80 551. *Charts, records, conservation game, etc.*
Royal Society for the Protection of Birds, Film Hire Library, 15 Beaconsfield Road, London NW1 2LE, tel. 01-451 1127. *16 mm films*

Saftvale, 11 Lord Street, Huddersfield HD1 1QA, tel. [0484] 40 131. *Models, posters, postcards of prehistoric animals*
Scottish Central Film Library, 74 Victoria Crescent Road, Glasgow G12 9JN, tel. 041-334 9314. *16 mm films, wide range*
Seminar Cassettes Ltd, 218 Sussex Gardens, London W2 3UD, tel. 01-262 7357. *Audio tapes*
Shell Education Service, Shell UK Ltd, PO Box 148, Shell-Mex House, Strand, London WC2R ODX, tel. 01-438 3000. *Film strips, charts, booklets*
Shell Film Library, 25 The Burroughs, London NW4 4AT, tel. 01-202 7803. *16 mm films, conservation, etc.*
Slide Centre, Dept T1, 143 Chatham Road, London SW11 6SR, tel. 01-223 3457. *35 mm slides*
Society for the Promotion of Nature Conservation, The Green, Nettleham, Lincoln LN2 2NR, tel. [0522] 52326. *Charts, leaflets, etc.*
Spiring Enterprises Ltd, North Holmwood, Dorking, Surrey RH4 5JL, tel. [0306] 3015. *Molecular models*
Studio Two Educational, 6 High Street, Barkway, Royston, Herts SG8 8EE, tel. [0763] 84759. *Slide/tape/record sets, materials on dinosaurs*

Tenovus Cancer Information Centre, 111 Cathedral Road, Cardiff CF1 9PH, tel. [0222] 42 851. *Anti-smoking posters*
Thomas Nelson & Sons Ltd, Lincoln Way, Windmill Road, Sunbury-on-Thames, Middx TW16 7HP, tel. Sunbury 85 681. *35 mm slides and OHPT* (*see Robert's book* Biology: a functional approach)

Resource lists

Transart Ltd, East Chadley Lane, Godmanchester, Cambridgeshire PE18 8AU, tel. [0480] 66999/0. *Overhead projector transparencies*

Unilever Education Section, PO Box 68, Unilever House, Blackfriars, London EC4P 4BQ, tel. 01-353 7474. *Charts, booklets*

United Nations Children's Fund, United Kingdom Committee, 46–48 Osnaburgh Street, London NW1 3PU, tel. 01-388 7487. *Posters, photographs, film strips, 35 mm slides (some only for loan)*

Visual Publications Ltd, 197 Kensington High Street, London W8 6BB, tel. 01-937 1568. *Film strips, multi-media materials*

Wellcome Film Library, The Wellcome Building, Euston Road, London NW1 2BP, tel. 01-387 4477. *Human parasites, diseases*

Woodmansterne Publications Ltd, Holywell Industrial Estate, Watford, Herts WD1 8RD, tel. Watford 28236. *35mm slides*

World Wildlife Fund, Education Project 1180, Brocklebank, Butts Lane, Woodmancote, Cheltenham, Glos GL52 4QH, tel. [024 267] 4839. *Multimedia materials on conservation*

World Wildlife Fund, 29 Greville Street, London EC1N 8AX, tel. 01-404 5691. *Charts, posters*

Zoological Society of London, Regent's Park, London NW1 4RY, tel. 01-722 3333. *Posters, information 'wheels' on zoo animals*

2 Suppliers : general scientific and biological

Ames Co., PO Box 37, Stoke Court, Stoke Poges, Slough SL2 4LY, tel. Farnham Common 2151. *'Albustix' and 'Clinistix' test sticks*

Andrew Stephens (1947) Co., 41 Dickson Road, Blackpool FY1 2AP, tel. [0253] 23 755. *Sphygmomanometers*

Aquatic Nurseries Ltd, Aqua House, Oak Avenue, Hampton, Middx TW12 3PR, tel. 01-979 6001/2 *and* 941 1313. *Cold-water, tropical and marine fish, aquaria and accessories*

Arnold E. J. & Son Ltd, Butterley Street, Leeds LS10 1AX, tel. [0532] 442 944. *General suppliers*

Arnold R. Horwell, 2 Grangeway, Kilburn High Road, London NW6 2BP, tel. 01-328 1551. *Range of biological laboratory supplies*

Associated Crates (Fabrications) Ltd, Coronation Street, Stockport, Cheshire SK5 7PL, tel. 061-480 3016. *Small mammal cages*

Baird & Tatlock (London) Ltd, Freshwater Road, Chadwell Heath, Essex RM1 1HA, tel. 01-590 7700. *General laboratory suppliers*

Baldwin, S. A., 32 Highfield Road, Purley, Surrey CR2 2JG, tel. 01-660 0629. *Palaeontological reproductions*

Bausch & Lomb UK Ltd, Highview House, Tattenham Crescent, Epsom Downs, Surrey KT18 5BR, tel. [07373] 60221. *Microscopes*

Baxter, Ronald N., 16 Bective Road, Forest Gate, London E7 ODP, tel. 01-534 6312. *Exotic entomological specimens*

BDH Chemicals Ltd, Poole, Dorset BH12 4NN, tel. [0202] 745 520. *Chemicals*

Beautiful Butterflies, High Street, Bourton-on-the-Water, Glos GL54 2AN, tel. [0451] 20 712. *Butterflies, stick insects and other exotic arthropods*

Biddolph, A., 'Moyne', Blacksole Bridge, Margate Road, Herne Bay, Kent CT6 6LA, tel. [022 73] 63073. *Dissecting instruments, lenses, etc.*

Bioscience, Harbour Estate, Sheerness, Kent ME12 1RZ, tel. [079 56] 67551. *Physiological equipment*

Bioserv Ltd, 38–42 Station Road, Worthing, West Sussex BN11 1JP, tel. [0903] 200844. *Biological supplies*

Bowman, E. K. Ltd, 12 Archway Close, London N19 3TD, tel. 01-272 1443. *Metal and plastic small mammal cages*

British Oxygen Co. Ltd, PO Box 17, Medical Works, Great West Road, Brentford, Middx TW8 9AL, tel. 01-560 3123. *Medical grade oxygen*

Cambrian Chemicals Ltd, Beddington Farm Road, Croydon CR0 4XB, tel. 01-686 3961. *Biochemicals*

Camlab Ltd, Nuffield Road, Cambridge CV4 1TH, tel. [0223] 62 222. *'Hach' water test kits and other laboratory items*

C & D (Scientific Instruments) Ltd, 439a London Road, Hemel Hempstead, Herts HP3 9BD, tel. [0442] 55194. *Microscopes*

Casella (London) Ltd, Regent House, Britannia Walk, London N1 7ND, tel. 01-253 8581. *Meteorological instruments*

Castle Associates, Rebourn House, North Street, Scarborough, North Yorkshire YO11 1DE, tel. [0723] 66347. *Sound level meters*

Charles Austen Pumps Ltd, 100 Royston Road, Byfleet, Surrey KT14 7PB, tel. Byfleet 43224/5. *Aquarium pumps*

Commonwealth Mycological Institute, Collection of Fungus Cultures, Ferry Lane, Kew, Surrey TW9 3AF, tel. 01-940 4086. *Fungi, other than pathogens, wood-rotting species and yeasts*

Cope & Cope Ltd, 57 Vastern Road, Reading, Berks RG1 8BX, tel. [0734] 54491. *'Cambridge' type mouse cage and others*

Corning Medical, St. Andrew's Works, Colchester Road, Halstead, Essex CO9 2DX, tel. [07874] 2461. *Colorimeters and pH meters*

Cossor, A. C. & Son (Surgical) Ltd, Accoson Works, Vale Road, London N4 1PS, tel. 01-800 1172/3. *Sphygmomanometers*

Culture Centre of Algae and Protozoa, 36 Storey's Way, Cambridge CB3 ODT, tel. [0223] 61378. *Algae and protozoa*

Denley Instruments Ltd, Daux Road, Billingshurst, West Sussex RH14 9SJ, tel. [0403] 81 3442. *Aluminium bottle and tube storage racks*

Ealing-Beck Ltd, 15 Greycaine Road, Watford, Herts WD2 4PW, tel. Watford 42261. *Microscopes, physiological equipment*
Edale Instruments (Cambridge) Ltd, Toft, Cambridge CB3 7RL, tel. [0220] 26 2782. *Thermistor thermometers*
Elliott Fox (Elstree) Ltd, Home Farm, Aldenham Road, Elstree, Herts WD6 3AY, tel. 01-953 5322. *Small mammal diets*

Gammaseed, 19 Royal Oak Drive, Bishops Wood, Staffs ST19 9AN, tel. [0785] 840 521. *Irradiated seed material*
GBI (Labs) Ltd, Shepley Industrial Estate, Audenshaw, Manchester M34 5DW, tel. 061-336 5418. *General biological suppliers*
Geological Laboratories, Lower Branscombe House, Ebford, Topsham, Exeter, Devon EX3 ORA, tel. [0392] 874260. *Geological specimens*
Gerrard Biological Centre, Worthing Road, East Preston, West Sussex BN16 1AS, tel. [090 62] 72071/5. *General biological suppliers*
Grant Instruments (Cambridge) Ltd, Barrington, Cambridge CB2 5QZ, tel. [0763] 60811. *Water baths, pumps, coolers*
Gregory, Bottley & Co., 30 Old Church Street, London SW3 5BY, tel. 01-352 5841. *Geological specimens*
Griffin & George Ltd, 285 Ealing Road, Alperton, Wembley, Middx HA0 1HJ, tel. 01-997 3344. *General laboratory suppliers*
Grundy (Teddington) Ltd, Somerset Works, Somerset Road, Teddington, Middx TW11 8TD, tel. 01-943 2441. *Food storage bins*

Haith, John E., Park Street, Cleethorpes, South Humberside DN35 7NF, tel. [0472] 57515/6. *Bird seed, grain, etc.*
Henley's Medical Supplies Ltd, Alexandra Works, Clarendon Road, London N8 0DL, tel. 01-888 3151. *Plastic taps and junction fittings*
Hopkins & Williams, PO Box 1, Chadwell Heath, Essex RM1 1HA, tel. 01-590 7700. *Chemicals*
Humex Ltd, 5 High Road, Byfleet, Weybridge, Surrey KT14 7QF, tel. Byfleet 51585. *Greenhouse equipment*

Irwin-Desman Ltd, 294 Purley Way, Croydon CR9 4QL, tel. 01-686 6441. *Ergometers, reaction timers, microscopes*

Jencons (Scientific) Ltd, Mark Road, Hemel Hempstead, Herts HP2 7DE, tel. [0442] 64641. *General laboratory suppliers*
Just Plastics, 5 Belgrave Gardens, London NW8 0QY, tel. 01-624 3826. *Plastic laboratory ware*

Kent Industrial Measurements Ltd, Hanworth Lane, Chertsey, Surrey KT16 9LF, tel. Chertsey 62671. *Oxygen and pH meters*
Koch Light Laboratories, Colnbrook, Berks SL3 0BZ, tel. [02812] 2262/5. *Organic chemicals, ATP, enzymes*

Larujon Locust Supplies, c/o Welsh Mountain Zoo, Colwyn Bay, Clwyd LL28 5UY, tel. [0492] 2938. *Locusts*
Luckham Ltd, Labro Works, Victoria Gardens, Burgess Hill, West Sussex RH15 9QN, tel. [044 46] 5348. *Aluminium bottle racks, etc.*

Marine Biological Association, The specimen department, The Laboratory, Citadel Hill, Plymouth, Devon PL1 2PB, tel. [0752] 21761. *Wide range of marine animals*
May & Baker Ltd, Dagenham, Essex RM10 7XS, tel. 01-592 3060. *Chemicals*
Medcalf Bros. Ltd, Cranbourne Road, Potters Bar, Herts EN6 3JN, tel. Potters Bar 56925. *Aquarium pumps*
Micro Instruments (Oxford) Ltd, 7 Little Clarendon Street, Oxford OX1 2HP, tel. [0865] 54466. *Microscope slides*
Millipore (UK) Ltd, Millipore House, Abbey Road, London NW10 7SP, tel. 01-965 9611. *Microbiological filtration equipment*
MRC Laboratory Animals Centre, Woodmansterne Road, Carshalton, Surrey SM5 4EF, tel. 01-643 8000. *List of accredited breeders and recognized suppliers*

Newbold & Bulford Ltd, Enbecco House, Carlton Park, Saxmundham, Suffolk IP17 2NL, tel. [0728] 2933. *Microscopes, 'Meopta' range*
Nordisk Diagonistics Ltd, 17 Halkingcroft, Langley, Berks SL3 7BB, tel. Slough 27767. *'Eldon' blood typing cards*
North Kent Plastic Cages Ltd, Home Gardens, Dartford, Kent DA1 1ER, tel. Dartford 21488/9 *and* 21188/9. *Small mammal cages*
Northern Biological Supplies, 3 Betts Avenue, Martlesham Heath, Ipswich, Suffolk 1P5 7HR, tel. [0473] 623 995. *Microscope slides*

Oertling, L & Co. Ltd, Cray Valley Works, St. Mary Cray, Orpington, Kent BR5 2HA, tel. Orpington 25771/6 *and* 36363/6. *Automatic balances*
Offord Scientific Equipment Ltd, 113 Lavender Hill, Tonbridge, Kent TN9 2AY, tel. [0732] 364 002. *Microscopes, Open University and 'Scientist', electronic thermometers, etc.*
Opax Ltd, 142 Silverdale Road, Tunbridge Wells, Kent TN4 9HU, tel. [0892] 25162. *Microscopes*
Optical Instrument Services (Croydon) Ltd, 166 Annerley Road, London SE20 8BD, tel. 01-778 7278. *Microscopes, 'Lumiscope', HSL, etc.*

Osmiroid Educational, E. S. Perry Ltd, Osmiroid Works, 104 Fareham Road, Gosport, Hants PO13 0AL, tel. [0329] 232 345. *Plastic quadrats, magnifiers, pollution kit, etc.*

Oxoid Ltd, Wade Road, Basingstoke, Hants RG24 0PW, tel. [0256] 61 144. *Microbiological media and small mammal diets*

Penlon Ltd, Radley Road, Abingdon, Berks OX14 3PH, tel. [0235] 24042. *'Longworth' small mammal traps*

Philip Harris Ltd, Lynn Lane, Shenstone, Staffs WS14 OEE, tel. [0543] 480 077. *General laboratory suppliers*

Philip Harris Biological Ltd, Oldmixon, Weston-super-Mare, Avon BS24 9BJ. tel. [0934] 413 063. *General biological suppliers*

Pitman Instruments, Jessamy Road, Weybridge, Surrey KT13 8LE, tel. Weybridge 46327/8. *Gas analysis pumps and detector tubes*

Portex Ltd, Hythe, Kent CT21 6JL, tel. [0303] 66863 *and* 60551. *Plastic tubing, connectors*

Practical Plant Genetics, 18 Harsfold Road, Rustington, West Sussex BN16 2QE. *Seeds for genetical investigations*

Prior WR Co. Ltd, London Road, Bishop's Stortford, Herts CM23 5NB, tel. [0279] 506414. *Microscopes*

Pyser-Britex (Swift) Ltd, Fircroft Way, Edenbridge, Kent TN8 6HL, tel. [0732] 864111. *Microscopes, 'Swift' range*

Raymond A. Lamb, 6 Sunbeam Road, London NW10 6JL, tel. 01-965 1834. *Chemicals for microscopy, etc.*

RS Components, PO Box 427, 13–17 Epworth Street, London EC2P 2HA, tel. 01-253 1222. *Electrical accessories*

Salcon Electronics, Springfield Road, Burnham-on-Crouch, Essex CM0 8TE, tel. [0621] 782268. *Egg incubators*

Scientific and Educational Aids, Vale Road, Windsor, Berks SL4 5JL, tel. Windsor 69218. *'Torbal' balances*

Scientific Instrument Centre, 1 Leeke Street, London WC1X 9JA, tel. 01-278 1581. *'Sartorius' balances*

Scientific and Research Instruments Ltd, Fircroft Way, Edenbridge, Kent TN8 6HE, tel. [0732] 864001. *Physiological equipment*

Shandon Southern Instruments Ltd, 93–95 Chadwick, Astmoor Industrial Estate, Runcorn, Cheshire WA7 1PR, tel. [092 85] 66611. *Chromatography equipment*

Simplex of Cambridge Ltd, Horticultural Division, Sawston, Cambridge CB2 4LJ, tel. [0223] 833 281. *Horticultural equipment*

Stanton Redcroft Ltd, PO Box 36, Orpington, Kent BR5 2ET, tel. Orpington 20555. *'Stanton' balances*

Sterilin Ltd, 43–45 Broad Street, Teddington, Middx TW11 8QZ, tel. 01-977 0944. *Plastic Petri dishes, etc.*

Stewart Plastics, Purley Way, Croydon, Surrey CR9 4HS, tel. 01-686 2231. *Range of plastic products, seed trays, etc.*

Sudbury Technical Products Ltd, Corwen, Clwyd LL21 0DR, tel. [0490] 2502. *Soil test kits*

Techne (Cambridge) Ltd, Duxford, Cambridge CB2 4PZ, tel. [0223] 832 401. *Water baths, coolers, pumps*

Thomas & Joseph Ltd, Castle House, Castle Meadow, Norwich, Norfolk NOR 41D, tel. [0603] 60331. *MS 222 anaesthetic for cold blooded vertebrates (ethyl m-aminobenzoate)*

Tintometer Ltd, The Colour Laboratory, Waterloo Road, Salisbury, Wilts ST1 2JY, tel. [0722] 27242/3/4. *Comparators*

Unilab Ltd, Clarendon Road, Blackburn BB1 9TA, tel. [0254] 57 643/4. *Environmental monitoring kits*

Vickers Instruments, Haxby Road, York YO3 7SD, tel. [0904] 24112. *Microscopes*

Visual Aids Centre, 78 High Holborn, London WC1V 6NB, tel. 01-242 6631 *and* 637 9837. *Materials for making visual aids*

Walden Precision Apparatus Ltd, Shire Hill, Saffron Walden, Essex CB11 3BD, tel. [0799] 3018 *and* 7104. *Environmental measuring equipment, chart recorder*

Watkins & Doncaster Ltd, Four Throws, Conghurst Lane, Hawkhurst, Kent TN18 5ED, tel. [058 05] 3133/4. *Insects, associated collecting and storage equipment*

Worldwide Butterflies Ltd, Over Compton, Sherborne, Dorset DT9 4QN, tel. [0935] 4608/9. *Insects, butterflies, moths, stick and leaf insects*

Xenopus Ltd, Holmesdale Nursery, Mid Street, South Nutfield, Redhill, Surrey RH1 4JY, tel. [073 782] 2687. *Amphibia and reptiles*

References

1 Advanced Biology Alternative Learning Project (1978), *ABAL working paper*, London: Inner London Education Authority
2 Ambercrombie, M. L. J. (1974), *Aims and techniques of group teaching*, 3rd ed, SRHE Monograph 12, Guildford: Society for Research into Higher Education
3 Andrewartha, H. G. (1961), *Introduction to the study of animal populations*, London: Methuen
4 Arkinstall, M. (1977), *Organizing school journeys*, London: Ward Lock
5 Arkless, S. M., Davies, N., Evans, D. E. J., and Hambleton, D. I. (1976), Combined science worksheets, Exeter: A. Wheaton (Years one and two and a Teacher's Guide)
6 Assistant Masters Association (1974), *Mixed ability teaching: a report of a survey conducted by the Association*, London: AMA
7 Barker, J. A., Hulme, S., and Smart, B. (1975a), *Breathing space*, People and Resources series and Teacher's Guide, London: Evans Brothers
8 Barker, J. A., Hulme, S., and Smart, B. (1975b), *Sink or swim?*, People and Resources series and Teacher's Guide, London: Evans Brothers
9 Barnes, D., Britton, J., Rosen, H., and the LATE (1971), *Language, the learner and the school*, rev. ed., Harmondsworth: Penguin
10 Barrass, R. (1979), 'Vocabulary for introductory courses in biology: necessary, unnecessary and misleading terms', *Journal of Biological Education*, vol. 13, no. 3, pp. 179-91 (Suggests a list of essential terms for biology up to 'O' level)
11 Beard, R. M. (1969), *An outline of Piaget's developmental psychology for students and teachers*, London: Routledge & Kegan Paul
12 Bennett, D. P., and Humphries, D. A. (1974), *Introduction to field biology*, 2nd ed., London: Edward Arnold (A standard work detailing techniques with short descriptions of specific habitats)
13 Biological Science Curriculum Study: An array of teaching materials for biology and related areas. Major biological publications include three main texts (blue, green and yellow versions), an advanced text and a remedial one. Available from various US publishers (details from BSCS, PO Box 930, Boulder, Colorado 80306, USA)
14 Bligh, D. A. (1972), *What's the use of lectures?* Harmondsworth: Penguin
15 Bracegirdle, B., and Miles, P. H. (1971-73), *An atlas of plant structure*, 2 vols., London: Heinemann Education
16 Bracegirdle, B., and Miles, P. H. (1978), *An atlas of chordate structure*, London: Heinemann Education
17 Bremner, J. (1962), 'Drawing and observation in biology lessons', *School Science Review*, vol. 43, no. 151, pp. 631-9

18 British Museum, Natural History (1978), *Introducing ecology: nature at work*, Cambridge: Cambridge University Press (Simple, largely pictorial introduction, produced in connection with exhibit at the British Museum, Natural History)
19 British Safety Council (n.d.), *First aid and emergency action*, London: British Safety Council (63-64 Chancellors Road, London W6 9RS)
20 Callender, P. (1969), *Programmed learning: its development and structure*, London: Longman
21 Carré, C. G. (1969), 'Audio-tutorials as adjuncts to formal lecturing in biology teaching at the tertiary level', *Journal of Biological Education*, vol. 3, no. 1, pp. 57-64
22 Carré, C. G., and Head, J. (1974), *Through the eyes of the pupil*, Science Teacher Education Project, London: McGraw-Hill (Collection of pupils' writing and illustration in science – provides an insight into the whys and wherefores of pupils' recording in science)
23 Carrick, T. (1977), 'A comparison of recently published textbooks for first examinations', *Journal of Biological Education*, vol. 11, no. 3, pp. 163-75 (Analyses of 12 texts published between 1974 and 1976, in terms of aims, content, presentation, etc.)
24 Carrick, T. (1978), 'Problems of assessing the readability of biology textbooks for first examination', *Journal of Biological Education*, vol. 12, no. 2, pp. 113-22 (Analyses, in terms of readability of 13 current texts, aspects of readability discussed)
25 Carter, G. (1979), *Handbook on environmental education in a totally urban setting*, Strasbourg: Council of Europe (Ideas about what has been done and what can be done in this topic)
26 Clark, E. (1973), *Fieldwork in biology: an environmental approach*, Basingstoke: Macmillan (Workbook with both secondhand evidence and practical investigations)
27 Cobb, V. (1974), *Science experiments you can eat*, Harmondsworth: Penguin
28 Cotton, J. (1979), 'Field teaching in inner cities: the William Curtis Ecological Park', *Journal of Biological Education*, vol. 13, no. 4, pp. 251-5
29 Darlington, A., and Brown, A. L. (1975), *One approach to ecology*, London: Longman (A simplified autecological approach)
30 Department of Education and Science (1972), *Safety in outdoor pursuits*, DES safety series no. 1, London: HMSO
31 Department of Education and Science (1976), *Safety in science laboratories*, DES safety series no. 2, London: HMSO
32 Department of Education and Science (1977), *The use of micro-organisms in schools*, Education pamphlet no. 61, London: HMSO
33 Eldin, H. L. (1968), *Know your broadleaves*, Forestry Commission booklet no. 20, London: HMSO
34 Eldin, H. L. (1970), *Know your conifers*, Forestry Commission booklet no. 15, London: HMSO
35 Evans, J. D. (1976), 'The treatment of technical vocabulary in textbooks of biology', *Journal of Biological Education*, vol. 10, no. 1, pp. 19-30 (Analysis of the vocabulary of six 'O' level biology texts)
36 Everett, K., and Jenkins, E.W. (1973), *A safety handbook for teachers*, London: John Murray

37 Fisher, R. A., and Yates, F. (1963), *Statistical tables for biological, agricultural and medical research*, 6th ed., London and Edinburgh: Oliver & Boyd
38 Ford, E. B. (1974), *Ecological genetics*, London: Methuen
39 Forestry Commission (1975), *Weeding in the forest*, London: HMSO
40 Freeman, W. H., and Bracegirdle, B. (1963), *An atlas of embryology*, London: Heinemann
41 Freeman, W.H., and Bracegirdle, B. (1966), *An atlas of histology*, London: Heinemann
42 Freeman, W. H., and Bracegirdle, B. (1971), *An atlas of invertebrate structure*, London: Heinemann
43 Freeman, W. H., and Bracegirdle, B. (1976), *An advanced atlas of histology*, London: Heinemann
44 Fry, P. (1968), 'Biological poster', letter in *The Times Educational Supplement*, 19 April 1968
45 Gagne, R. M. (1970), *Conditions for learning*, 2nd ed., London: Holt Rinehart & Winston
46 Gardner, B. (compiler) (1976), (1964), *Up the line to death: the war poets 1914–18, an anthology*, rev. ed., London: Methuen (Contains Wilfred Owen's poem 'Dulce et Decorum est')
47 George, C., (ed.) (1976), *Nonstreamed science: organisation and practice*, Study series no. 10. Hatfield, Association for Science Education (Collection of articles based on the authors' personal experiences with non-streamed classes)
48 Gilliland, J. (1972), *Readability*, Sevenoaks: Hodder & Stoughton (The concept of readability and how it can be assessed)
49 Gilman, D. (1977), *Urban ecology*, London: Macdonald (This pupils' book provides a short, simple and well illustrated introduction)
50 Graham, W. (1978), 'Readability and science textbooks', *School Science Review*, vol. 59, no. 208, pp. 545–50 (Brief introduction to readability tests)
51 Green, E. (ed.) (1976), *Towards independent learning in science*, St Albans: Hart-Davis (Variety of contributions ranging over school and higher education)
52 Hambler, D. J., and Field, D. (1969), 'Views on practical notebooks in the teaching of biology', *Journal of Biological Education*, vol. 3, no. 1, pp. 75–90
53 Hammond, R. A., and Roach, D. K. (1978), 'Progress and evolution of a self-study zoology course at University College Cardiff', *Journal of Biological Education*, vol. 12, no. 2, pp. 123–32
54 Heath, D. J. (1975), 'Coloured beads for genetic modelling', *Journal of Biological Education*, vol. 9, no. 2, pp. 71–4
55 HMSO (1974), *Health and Safety at Work etc. Act*, London: HMSO
56 Hutt, C. (1972), *Males and females*, Harmondsworth: Penguin
57 Inexpensive Science Teaching Equipment Project (1972), *Guidebook to constructing inexpensive science teaching equipment*, vol. 1, *Biology*, Maryland: Science Teaching Centre, University of Maryland
58 Jenkins, P. F. (1973), *School grounds: some ecological enquiries*, London: Heinemann (Simple approach to ecological work in the school environment)
59 Kelly, A. V. (1975), *Case studies in mixed ability teaching*, London: Harper and Row (Useful background on mixed ability teaching in practice)

60 Kelly, A. V. (1978), *Mixed ability grouping, theory and practice*, London: Harper and Row (Useful general introduction)
61 LAMP Project (Least academically motivated pupils in the secondary school): A series of topic briefs produced by groups of science teachers and published by the Association for Science Education
62 Limbird, J. E., and Brogden, E. A. (1968), 'Biological poser: embryo in class', letter in *The Times Educational Supplement*, 5 April 1968
63 Lunzer, E. and Gardener, K. (1979), *Effective use of reading*, London: Heinemann (Report of the Schools Council project of reading. See particularly sections on textbooks and reading in science)
64 Mackean, D. G. (1973), *Introduction to biology*, 5th ed. London: John Murray
65 Markle, S. M. and Tiemann, P.W. (1970), 'Problems of conceptual learning', *Journal of Educational Technology*, vol. 1, no. 1, pp. 52–62
66 Marland, M., *et al.* (1977), *Language across the curriculum, the implementation of the Bullock Report in the secondary school*, London: Heinemann (Various contributions concerned with the implementation of the Bullock Report in schools)
67 Marshall, D., and Tranter, J. (eds.) (1978), Ecopacks: (1) *Woodlice*, (2) *Air pollution*, (3) *Trees*, (4) *Water pollution*, (5) *Trapping*, (6) *Soil organisms*, Basingstoke: Globe Education, in association with Inner London Education Authority (Series of units devised by biology teachers from ILEA schools. Units contain a variety of teaching materials)
68 Martin, W. K. (1965), *The concise British flora in colour*, London: Ebury Press and Michael Joseph
69 McDiarmid, A. (1975), 'Pets and the zoonoses' in P. J. Kelly and J. D. Wray *The educational use of living organisms – a source book*, London: English Universities Press
70 McPhail, P. (1972), *Consequences*, London: Longman (part of the Schools Council's Moral Education Project)
71 McPhail, P. *et al.* (1972), *Lifeline*, London: Longman (Schools Council Moral Education Project 13–16 years resource material and teachers' guide)
72 Medawar, P. B. (1972), *The hope of progress*, London: Methuen
73 Moss, S., and Theobald, D. (1979), 'An answer to the teaching of ecology', *Journal of Biological Education*, vol. 13, no. 1, pp. 17–24 (Simple investigations used in Africa, but more widely applicable)
74 Mullin, B., (1966), 'The working of the human eye', pt. 11, *Physics Education*, vol. 1, no. 2, pp. 104–6
75 Nuffield Foundation (1970a), *Combined Science, Teachers' Guide*, 3 vols., London and Harmondsworth: Longman and Penguin
76 Nuffield Foundation (1970b), Nuffield Advanced Biology, *Maintenance of the organism, a laboratory guide; Organisms and populations, a laboratory guide; The developing organism, a laboratory guide; Control and co-ordination in organisms, a laboratory guide; Teachers' Guides 1 and 2; Laboratory book: a technical guide* (1971), Harmondsworth, Penguin
77 Nuffield Foundation (1970c), Nuffield Advanced Biology, *Study guide: evidence and deduction in biological science*, Harmondsworth: Penguin (students' book and a teachers' guide)

78 Nuffield Foundation (1971), Nuffield Secondary Science, *Theme 1, Interdependence of living things; Theme 2, Continuity of life; Theme 3, Biology of man; Theme 4, Harnessing energy; Theme 5, Extension of sense perception; Theme 6, Movement; Theme 7, Using materials; Theme 8, The earth and its place in the universe; Teachers' Guide; Apparatus Guide* (1972); *Examining at CSE level* (1972), London: Longman

79 Nuffield Foundation (1974-5), Revised Nuffield Biology, *Text 1, Introducing living things* (1974); *Text 2, Living things in action* (1975); *Text 3, Living things and their environment* (1975); *Text 4, The perpetuation of life* (1975); and corresponding *Teachers' Guides*, London: Longman

80 Nuffield Foundation (1975), Science Teaching Project: Chemistry (rev. ed.) A. Mansell, *The Halogens* (Study sheets), London: Longman (Wilfred Owen, 'Dulce et Decorum est)

81 Open University, language and learning course team (1972), *Language in Education: a source book*, London: Routledge & Kegan Paul (See, in particular, the section on language in the classroom)

82 Postlewait, S. N., et al. (1976), *Minicourses in biology*, London and Philadelphia: W.B. Saunders (A series comprising study guides; audio tapes; visual materials and instructor's manuals)

83 Pringle, M. L. K. (1970), *Able misfits*, London: Longman

84 Pringle, M. L. K. (1980), *The needs of children*, 2nd ed., London: Hutchinson

85 Pringle, M. L. K. (1973), *The roots of vandalism and violence*, London: National Children's Bureau

86 Reid, D. (1973), 'Independent learning in science – a review of progress made, with some suggestions for future development', in E. L. Green (ed.), *Individual and small group methods in the teaching of science. A report on the course/conference held at Countesthorpe College 17-19 April 1973*

87 Reid, D. J., and Booth, P. (1969), 'The use of individual learning with the Nuffield biology course', *School Science Review*, vol. 50, no. 172, pp. 493-506

88 Reid, D. J., and Booth, P. (1971), 'Independent learning in biology – results of trials, 1968-70', *School Science Review*, vol. 52, no. 180, pp. 500-12

89 Reid, D., and Booth, P. (1971-5), Biology for the individual: *Introduction to the series* (1971); (1) *Sorting animals and plants into groups* (1971); (2) *How life begins* (1971); (3) *Movement in animals* (1971); (4) *Support in animals and plants* (1972); (5) *Patterns of life in hot, cold, and dry climates* (1972); (6) *Working with very small things* (1974); (7) *War against disease* (1974); (8) *Plant reproduction* (1973), London: Heinemann

90 Reid, D. (1978-9), Biology for the individual: (9) *Skeleton and muscles* (1978); (10) *Food, feeding and digestion* (1979), London: Heinemann

91 Reid, D., and Booth, P. (1974), 'Independent learning', in C. Selmes (ed.), *New movements in the study and teaching of biology*, London: Temple Smith (Simple, concise introduction to the value of independent learning in schools)

92 Report of a working party of the Biological Education committee of the Royal Society and the Institute of Biology (1975), 'The dissection of animals in schools', *Journal of Biological Education*, vol. 9, nos. 3/4, pp. 146-54

93 Report of the Central Advisory Council for Education (England), A (Newsom Report) (1963), *Half our future*, London: HMSO

94 Report of the committee of inquiry appointed by the Secretary of State for Education and Science under the Chairmanship of Sir Alan Bullock, FBA (1975), *A language for life,* London: HMSO
95 Schools Council (1964), Working paper no. 1: *Science for the young school leaver,* London: HMSO
96 Schools Council (1972), *Out and about: a teacher's guide to safety on educational visits,* London: Evans/Methuen
97 Schools Council (1973-5), Schools Council Integrated Science Project: (1) *Building Blocks;* (2) *Interactions and change;* (3) *Energy:* (4) *Interactions and change,* London: Longman (Range of publications, pupils' manuals, teachers' and technicians' guides)
98 Schools Council (1974-7), Educational Use of Living Organisms Project: J. D. Wray (1974a), *Animal accommodation in schools;* J. D. Wray (1974b), *Small mammals;* P. J. Kelly and J. D. Wray (1975), *The educational use of living organisms - a source book;* L. C. Comber (1976), *Organisms for genetics;* G. D. Bingham (1977), *Plants;* P. Fry (1977), *Micro-organisms,* Sevenoaks: Hodder & Stoughton
99 Spencer, J. (1977), 'Games and simulations for science teaching', *School Science Review,* vol. 58, no. 204, pp. 397-413
100 Stamp, R. D. and Harrison, W. (1975), 'The great blood race', in *Science games one, biology and general science,* London: Longman
101 Sturges, L. W. (1975), *Non-streamed science: A teacher's guide,* Study Series 7, Hatfield: Association for Science Education (Good basic material and sources for further reading, planning, methods, materials, assessment, exceptional pupils, etc.)
102 Sutton, C. (ed.) (1979), 'Written work in science lessons: annotated extracts', Occasional paper, Science Education series, Leicester: University of Leicester, School of Education
103 Tawney, R. H. (1966), *The radical tradition,* Harmondsworth: Penguin
104 Taylor, J., and Walford, R. (1978), *Learning and the simulation game,* Milton Keynes: The Open University Press
105 Tinbergen, D., and Thornburn, P. (1976), *Integrated Science,* The Wreake Valley Project Books 1, 2 and 3, London: Edward Arnold
106 Tinbergen, N. (1977), 'On war and peace in animals and man', in T. E. McGill, *Readings in animal behaviour,* 3rd ed., London: Holt Rinehart & Winston
107 Tribe, M. A., Eraut, M. R., and Snook, R. K. (1975), *Tutors' guide,* Cambridge: Cambridge University Press
108 Tricker, B. J. K., and Dowdeswell, W. H. (1970), Nuffield Advanced Science: *Projects in Biological Science,* Harmondsworth: Penguin
109 United Nations Educational, Scientific and Cultural Organisation (1962), *UNESCO source book for science teaching,* Paris: UNESCO
110 Uttley, A. (1969), *A country child,* Harmondsworth: Puffin
111 Walford, R. (1969), *Games in geography,* London: Longman
112 Watson, J. D. (1968), *The double helix,* London: Weidenfeld and Nicolson
113 Wilson, R. W. and Wright, D. F. (1972), *A field approach to biology* (Teachers' Guide and 4 student books), London: Heinemann (Contains detailed suggestions for work based on a problem-solving approach)

114 Wray, J., and the Schools Council Science Committee Working Party on dissection and experimentation in schools (1974), *Recommended practice for schools relating to the use of living organisms and material of living origin*, London: English Universities Press
115 Writing across the curriculum 11-13 project (1975), *Writing in science: papers from a seminar with science teacher*, London: Schools Council/London Institute of Education

Audio-visual aids

116 *Air pollution*, Introductory study pack M.80505/8, Philip Harris Biological Ltd (For five students, larger pack for ten available)
117 'Banta' viewers, Banta (Educational suppliers), 279 Church Road, London SE19 2QQ
118 *Blood circulation*, Macmillan (Standard or super 8 mm film loop)
119 *Breathing*, Macmillan (Standard or super 8 mm film loop)
120 *Breeding of roses, The*, Longman (Standard 8 mm film loop)
121 *Capillary circulation of blood*, Longman (Standard 8 mm film loop)
122 *Cleaning mechanism of the lungs*, Longman (Standard 8 mm film loop)
123 *Emergency resuscitation: mouth to mouth* (CEM/NA/E1); *mouth to nose* (CEM/NA/E2), Eothen International Ltd (Super 8 mm film loops)
124 *Fertilization in the marine worm* Pomatoceros triqueter, Longman (Super 8 mm film loop)
125 *First days of life*, Les Films du Levant (in collaboration with Guiqoz-France) (16 mm film, colour, sound, 22 minutes) (1972). Available on sale from: Boulton-Hawker Films Ltd; *or*, on hire from: NAVAL, Concord Films Council and Scottish Central Film Library
126 Food exchange playing cards, British Diabetic Association, 10 Queen Anne Street, London W1M 0BD
127 *Geological time scale*, parts 1 and 2, Longman (Standard 8 mm film loops)
128 Harris/Gene-kit, M. 89100/5, Philip Harris Biological Ltd
129 *Heart in action, The*, Macmillan (Standard or super 8 mm film loop)
130 *Laboratory precautions*, Longman (Standard 8 mm film loop)
131 *Lice biting and sucking*, M.39501/4 (a set of 10, 35 mm colour slides). Available from: Philip Harris Biological Ltd
132 *Life history of the Cabbage White butterfly*, Oxford Scientific Films Ltd (16 mm film, colour, sound, 15 minutes) (1968). Available on hire from: NAVAL
133 *Measuring the very small*, Longman (Standard and super 8 mm film loop)
134 *Progeny testing*, Longman (Standard 8 mm film loop)
135 *Results of the selective breeding of two varieties of hen, The*, Longman (Standard 8 mm film loop)
136 *Stream and river pollution*, Introductory study pack M.80525/3 Philip Harris Biological Ltd (For five students, larger pack for ten available)
137 *Streamline flow in liquids*, S.64, A. M. Lock & Co. Ltd, Neville Street, Middleton Road, Oldham OL9 6LF, tel. 061-024 0333
138 *War to the last itch* (16 mm film, colour, sound, 16 minutes (1963). Available on hire from: Peter Darvill Associates Ltd

139 *Windows of the soul* (16 mm film, colour, sound, long verson 62 minutes, short version 28 minutes). Available on hire from: Fact and Faith Films, 37 Coton Road, Nuneaton, Warws CV11 5TW

Further information

Books

Archenhold, W.F., Jenkins, E.W. and Wood-Robinson, C. (1977), *Addresses for science teachers,* Leeds: Centre for Studies in Science Education, University of Leeds
Bremner, J. (1967), *Teaching biology,* London: Macmillan (Brief, but broad coverage of the major areas involved in the teaching of biology)
DELTA (Directory of environmental literature and teaching aids), Council for Environmental Education (24 London Road, Reading RG1 5AG)
Department of Education and Science (1979), *The environment sources of information for teachers,* London: DES
Dixon, A. (ed.) (1977), *Useful addresses for biologists*, Hatfield: Association for Science Education
Falk, D. F. (1971), *Biology teaching methods,* London: John Wiley (American publication, particularly concerned with teaching strategies and tactics, rather than consideration of the teaching of specific topics)
Foster, D. (1979), *Resource-based learning,* Study series no. 14, Hatfield: Association for Science Education (Management and planning of resources, collection, planning and retrieval)
Kramer, L. M. J. (1975), *Teaching the life sciences,* London: Macmillan (Detailed account, with particularly useful reference lists)
Maslin, D. (ed.) (1978), *Biological science: a subject index of audio-visual materials,* London: Institute of Biology (Lists under topic major AVA materials of value in biology teaching at all levels)
Mayer, W. V. (ed.) (1978), *Biology teachers' handbook,* 3rd ed., Chichester: John Wiley (Although basically concerned with the BSCS materials it contains many valuable ideas. See 'Invitations to enquiry' as an approach to the use of secondhand evidence)
Sutton, C. R., and Haysom, J. T. (1974), *Art of the science teacher,* Science teacher education project, London: McGraw-Hill (Although not much specifically biological, a mine of valuable ideas)

Journals

American Biology Teacher, National Association of Biology Teachers, 11250 Roger Beacon Drive, Reston Va. 22090, USA (Published eight times a year)

Biology and Human Affairs, British Social Biology Council, 69 Eccleston Square, London SW1V 1PJ (Published twice a year)

Education in Science, Association for Science Education, College Lane, Hatfield, Herts AL10 9AA (Published five times a year)

Journal of Biological Education, Institute of Biology, 41 Queen's Gate, London SW7 5HU (Published four times a year)

Natural Science in Schools, School Natural Science Society, 5 Upper Park Road, Kingston upon Thames, Surrey KT2 5LB (Published three times a year)

New Scientist, King's Reach Tower, Stamford Street, London SE1 9LS (Published weekly)

School Science Review, Association for Science Education, College Lane, Hatfield, Herts AL10 9AA (Published four times a year)

Scientific American, 415 Madison Avenue, New York, NY 10027, USA (published monthly)

Index

ability
  drawing, 57–8
  lack of, 121–2
  mixed, 40–65
  reading, 54–5
  technical, 57
  verbal, 47–8
abstraction, 45
aesthetic education, 80, 83
animals
  found dead, 126
  use of, 124–7
apparatus, simple, 38
application, 27, 94
assessment, 61
  of course work, 61–5
audio-visual aids, list of suppliers, 155–61

behaviour, human, 38, 138–47
behavioural objectives, 31, 122
biological suppliers, 161–6
birds, use of, 134
blood system, 25–7, 110

cemetery, use of, 134–5
circus, 93–6
colonization, 33, 34
communication, 47–54
  non-verbal, 8
  visual, 99–100
comprehension, 55–6
concept, 31, 94–5, 131–3
consolidation, 27–8
course work assessment, 61–5
crossword puzzles, 106, 107

discipline, 6–18
discussion, learning by, 72–8, 98–9
dissection, 127–8, 130
drawing, 56, 57–8, 78–83, 99–100
  in examinations, 116

ecology, 132–4
embryos, use of, 125–6
emotional education, 91
energy transfer, 93
English, not first language, 8
evaluation, 28
examination
  practical, 115–6
  written, 114–5

fact files, 4
factual objectives, 122
field work, 130–5
film, use of, 33, 69, 86
film loop, use of, 23–4, 87
film-strip, use of, 84
Flora Game, 108, 109
food tests, 69

games, 106–10
generalization, 45
Giky Martables, 7
graphs, 51–2
Great Blood Race, 110
group discussion, 98–9
group work, 5, 9, 12, 49, 95–6, 98–9

'Happy Families', game based on, 108, 109, 132
historical approach, use of, 136–7
human behaviour, 38, 138–46
human body, use of, 33, 35–8, 138–46

independent learning, 66-7, 100, 101, 105, 113-14, 153-4

kidney action, model of, 38-9
kitchen, use of, 39

laboratory, teaching without, 71-2
language, 7-8, 29-30
learning
  independent, 66-7, 100, 101, 105, 113-4, 153-4
  programmed, 67-8
learning pace, 41
lecturing, 116-18
  notes, 117
lesson planning, 4, 11, 18, 20-39
  problems and solutions, 33-9
lesson sequencing, 21-4
lesson timing, 21
lessons, revision 31-2
library, use of, 54-5, 114
lungs, ventilation of, 21-4

Malpighi, Marcello, 23, 136-7
mathematical symbolism, 49-50
measurement of humans, 35, 38, 138-46
memory
  long-term, 27, 46
  short-term, 44
microscope, use of, 68, 80
mixed ability teaching, 40-65
model, 38-9, 52-3, 84-6, 89-90
motivation, 28-9, 59-60, 87

Newsom pupil, 45-6
nomogram, use of, 51-2
non-verbal communication, 8, 37-8
note-taking, 112-13

objectives
  behavioural, 31, 122
  factual, 122
osmosis, 52, 86, 93
overhead projector, use of, 33, 74, 84, 92

perception, training in, 78-80, 101, 103
pictorial symbolism, 51-2, 80
pig's trotter, use of, 127-8

practical work, 2, 68-70
programmed learning, 67-8
project work, 83
provocation, attention-getting, 13-14, 123
pupils
  as teachers, 110-12, 113
  16 to 18-plus, 112-18
  what worries, 18-19

Quest, 47-8, 57, 101, 104, 151-2

readability, 29-30
record of work, 10, 28, 62-3, 97, 99
registration, 11-12
resources, teaching without, 33-9
respect, 14-18
rivers, use of, 134
revision, 28, 31-2

safety, 69, 129-30
Schools Council Moral Education Project, 129
Schools Council Integrated Science Project, 46
seminars, 113
senses, 101, 102
sensitivity of pupils, 91, 126
skills, 54-8, 151-2
slides, 35 mm, 84
'Snap', game based on, 107
stimuli, sensitivity to, 101, 102
subcultures, different, 58-60
suppliers
  audio-visual aids, 155-61
  general scientific and biological, 161-6
symbolism
  mathematical, 49-50
  pictorial, 51-2, 80
  realistic, 52-3

tape recorder, 8, 87
tape/slide programme, 87
taxonomy, 108, 109, 132, 133
teach, impossible to, 122-3
teaching
  facts, 4
  making mistakes, 4-5

Index

teaching practice, 3, 6–7, 15–16, 18–19
television, educational, 87–8, 114
terminology, technical, 7–8
textbooks, 29–30, 116
thermometers, clinical, 38–9, 57, 152
thought, logical, 56–7
tree, use of, 134

variation, 24, 31
ventilation of human lungs, 21–4
visual aids, 83–93
  use of, 88–91
voice, 8, 9, 15–16

word bank, 29
work record, 10, 28, 62–3, 97, 99
workcard, 8, 100, 101, 127; *see also* worksheet
worksheet, 4, 8, 34, 36, 37, 44, 48, 50, 72, 96–106, 136–54
  graded, 44
writing, 30, 99
writingboard, 91–3

xeromorphs, 134